Building Ontologies with Basic Formal Ontology

Building Ontologies with Basic Formal Ontology

Robert Arp, Barry Smith, and Andrew D. Spear

The MIT Press
Cambridge, Massachusetts
London, England

This book was set in Stone Sans and Stone Serif by Toppan Best-set Premedia Limited.

Library of Congress Cataloging-in-Publication Data is available.

ISBN: 978-0-262-52781-1 (paperback)

To Susan, Sandra, and Maria Teresa

Contents

Contents

Preface

In recent decades we have seen the gradual expansion of the use of computers and computing technology in all areas of human life. In the sciences the promise of computers to store, manage, and integrate tremendous amounts of data and information has given rise to new disciplines focused on data and information, and to new interdisciplinary fields such as biomedical informatics, materials informatics, geospatial informatics, and many more.

One increasingly dominant strategy for the organization of scientific information about the world in computer-friendly form is associated with the term "ontology" (or sometimes "ontological engineering" or "ontology technology" or "applied ontology"), understood as meaning (roughly) a controlled vocabulary for representing the types of entities in a given domain.

This strategy has been most conspicuously successful in the field of biology and biomedicine, where its proponents have come to view the task of organizing scientific information as requiring an unusually broad collaboration involving not only computer and information scientists, biologists, and clinicians, but also linguists, logicians, and, occasionally, *philosophers* interested in the study of the basic categories of reality. This book is an introduction to the field of applied ontology thus conceived. It explains the needs which ontologies have been designed to meet, explains what an ontology itself is, and outlines in detail principles of best practice for approaching the task of ontology design. The book also outlines a specific formal or top-level ontology, the Basic Formal Ontology, and provides illustrations of its use.

All three coauthors of this work were trained as philosophers, though all have become involved, in different ways, in applied ontology projects in biomedicine and related fields. All share the belief that philosophical ideas and theories can play an important role in advancing the quality of work in ontological applications, and the chapters that follow are very much a product of this belief. We have used philosophical ideas throughout—though our philosophical colleagues will say that we have sometimes done so incautiously, and certainly with what is for their purposes insufficient detail. What follows is not, however, intended as a contribution to philosophy. It is intended, rather, to form part of what we conceive as the rich, new technical discipline of ontology.

Acknowledgments

This book has been a long time in the making and has benefited from the collaboration and critical comments of numerous individuals. Andrew Spear wrote a first rough draft of the manuscript under the direction of Barry Smith at the Institute for Formal Ontology and Biomedical Information Science (IFOMIS) in Saarbrücken, Germany, in 2006. In 2007 Robert Arp became involved in the project under the auspices of the National Center for Biomedical Ontology, leading to substantial revision and expansion of the manuscript in a collaborative effort of over eight years. The result is the current book, which bears equally the stamp of each of us, and of our limitations.

After so many years of discussion and input, it would be impossible to recognize everyone who has made some contribution to this manuscript here and to the development of Basic Formal Ontology (BFO) on which it is based. However, we would like to acknowledge the comments and critical advice of Mauricio Almeida, Jonathan Bona, Mathias Brochhausen, Roberto Casati, Werner Ceusters, Melanie Courtot, Lindsay Cowell, Randall Dipert, William Duncan, Bastian Fischer, Albert Goldfain, Pierre Grenon, Janna Hastings, Boris Hennig, William Hogan, Leonard Jacuzzo, Ingvar Johansson, Waclaw Kusnierczyk, Kristl Laux, Richard Lee, Tatiana Malyuta, William Mandrick, Kevin Mulligan, Chris Mungall, Darren Natale, Fabian Neuhaus, Snezana Nikolic, Chris Partridge, Bjoern Peters, Anthony Petosa, Mark Ressler, Robert Rovetto, Ronald Rudnicki, Alan Ruttenberg, Emilio Sanfilippo, Richard Scheuermann, James Schoening, Yonatan Schreiber, Stefan Schulz, Ulf Schwarz, Selja Seppälä, Shane Sicienski, Peter Simons, Holger Stenzhorn, Kerry Trentelman, Achille Varzi, and Jie Zheng. Naturally, they bear no responsibility for the many shortcomings that remain.

Chapters 5, 6, and parts of 7 are based on the draft specification of BFO 2.0, which contains also formal definitions of the terms introduced in these chapters as well as associated axioms and theorems and considerable further explanatory material. Many of the persons mentioned above provided invaluable assistance in creating this specification, but we wish to mention especially Werner Ceusters and Alan Ruttenberg.

We are also particularly grateful to Ingvar Johansson, Waclaw Kusnierczyk, and Sne-zana Nikolic, who read and commented on early drafts of the whole document, and to Mark Musen, Director of the National Center for Biomedical Ontology, which supported our work on the preparation of these early drafts. We are grateful also to the National Institute for Human Genome Research, the Alexander von Humboldt Foundation, the Volkswagen Foundation, and the European Union, which provided funding for this work. None of these organizations is responsible in any way for the content of what follows.

Introduction

Overwhelmed with Information

Today more than ever before in history, we live in an age of information-driven science. In all areas of the life sciences, in particular, well-organized and well-funded research groups are carrying out sustained and systematic research into areas of fundamental biological concern, yielding ever-larger quantities of information that is accessible only with the aid of computers. Vast amounts of information are being produced daily as a result of new types of high-throughput technology in areas such as next generation sequencing, molecular screening, and 2-, 3-, and 4-D imaging at multiple scales from molecules through cells and cell populations up to whole brains. At the same time the contents of scientific journals are increasingly being made available in forms that make them accessible to automated search and processing.

Already, the sheer quantity of available scientific information is becoming overwhelming, and this new information can be used effectively only if there is some strategy for ensuring its progressive integration with the information already existing, and for making it readily available in formats understandable to both computers and to human beings. The progress of science requires that the results being achieved in Pittsburgh or Berkeley should be able to build on the results already achieved in Peking or Bangalore. For these reasons scientific information needs to be stored, standardized, processed, and made available in a way that overcomes the idiosyncrasies of particular research groups and technologies.

Computers are able to store massive amounts of information, more than any single human memory could possibly retain, and they are able to retrieve specific information in focused ways, and to perform logical operations—to reason, in a sense—across this information in ways that go beyond what would be achievable by any human mind. It is thus by now self-evident that the problems of storing, organizing, retrieving, integrating, and making available the ever-growing quantity of scientific information must be met by exploiting the powers of computers. Indeed, many are now pursuing a vision of the future in which the knowledge possessed by experts working in the same or related fields would be organized and stored in interconnected computer repositories in such a way that it would become accessible to any human being or to any suitably equipped computer anywhere in the world, in real time, and be continuously updated in light of new results.

For this to occur, however, databases must be created in such a way that their contents can be shared and used in equally effective ways not only by those who have created them, and by those who populate and maintain them, but also by as yet unidentified populations of external users. For reasons that we will discuss, this goal has not yet been attained. A recent survey of healthcare data scientists finds that it is the diversity of data available, not the quantity, that poses the primary challenge in making use of electronic medical data and information.[1] Similarly, a recent report from the Office of the National Coordinator for Health Information Technology points out that "despite significant progress in establishing standards and services to support health information exchange and interoperability, it is not the norm that electronic health information is shared beyond groups of health care providers who subscribe to specific services or organizations."[2]

This same point was reiterated in expert witness testimony before the U.S. Congress in July 2014.[3] Electronic information needs to be interoperable, shareable, and reusable. Think of a doctor specializing in a certain rare disease with immediate access to the most current information about all the patients suffering from this disease and about novel treatment outcomes. Imagine, still more ambitiously, a single, integrated biomedical knowledge base, a kind of Great Biomedical Encyclopedia, comprehending all biomedical knowledge within one constantly evolving system. Such possibilities are not beyond our reach. Experiments are currently being made under headings such as "semantically enhanced publishing" or "Big Data to Knowledge" (BD2K), and the potential benefits of success of such ventures are easy to appreciate. These benefits can be achieved, however, only through radical improvements in the degree to which the information systems involved are capable of interoperation.

What is needed in order to advance interoperability in a way that would address the shortcomings just described? First, consistency in the way standard sorts of data are described—consistency in units of measure, in terms for chemical, biological, and clinical entities of different sorts, in tagging of sequence and image data, and many more. Second, however, it requires consistency at the level of scientific assertions: when one biologist identifies a certain enzyme as exercising a function of a certain type when in a certain sort of cellular location, then his assertions have to be comparable to—and logically combinable with—the assertions of other biologists who have identified, for instance, that enzymes of that sort interact with certain other proteins whenever such a function is realized. The task of collecting and organizing scientific data in a way that supports such comparisons and combinations in the era of information-driven science is, it turns out, much more difficult than was anticipated in the era when scientific reasoning was carried out by human beings without the aid of computers. In the pages that follow we address some of the obstacles to interoperability and attempt to show how ontology can help overcome them.

Given the continuing and accelerating developments in science and information-based technology, a perspective geared toward enhanced accessibility of databases is necessary, especially in highly data-intensive and rapidly evolving fields such as biomedicine. The attempt to achieve these ends is being carried out in the work of many current institutions and initiatives, including the Gene, Cell and Protein Ontology Consortia, the ISA Commons, the Neuroscience Information Framework (NIF) Standards, and the cROP (Common Reference Ontologies for Plants) and OBO Foundry (Open Biological and Biomedical Ontologies Foundry; formerly Open Biomedical Ontologies Foundry) initiatives. In addition, the National Institutes of Health and its counterparts in other countries are providing significant grant support to researchers in the hope of making information resulting from biomedical research more easily accessible through new publishing policies and through the use of new types of unified information systems.

Obstacles to Accessibility: Human and Technical Idiosyncrasy

As we have already indicated, there are a number of obstacles to achieving effective accessibility, interoperability, and reusability of data and information. The first is that the members of the community of scientific (including clinical) researchers use different, and sometimes incommensurable, specialist terminologies and formats in describing the results of their research. The second is that they also use a variety of different computer technologies to encode and store their results, in part because these technologies are themselves rapidly evolving; in part because software and database developers have incentives to create new artifacts rather than to reuse what already exists. These two issues can be labeled the *human idiosyncrasy* and the *technological idiosyncrasy* problems, respectively. Different researchers encode their research using different terminologies and coding systems, and using different computing formats and software.

Table 0.1
Labels for C-C motif chemokine 15 in different research communities

CCL15	mrp-2b	mip-1(delta)	ncc3	CC motif	mip-related protein
ccl-15	MIP5	mip1(delta)	NCC3	chemokine 15	macrophage
HCC-2	lkn-1	mip1delta	ncc-3	c-c motif	inflammatory protein 5
hcc-2	MIP-5	mip-1-delta	SY15	chemokine 15	macrophage
hcc2	mip5	mip-1delta	SCYL3	chemokine CC2	inflammatory protein-5
hmrp-2b	mip-5	mip1d	scyl3	chemokine CC-2	new CC chemokine 3
HMRP-2B	MIP-1D	MRP-2B	scya15	chemokine cc-2	small ccl-15
LKN1	mip-1d	Mrp-2b	SCYA15	chemokine (c-c	small inducible
LKN-1	MIP-1	NCC-3		motif) ligand 15	cytokine A15
	delta			inducible cytokine	small-inducible
				subfamily A	cytokine a15
				(CysCys), member 15	small-inducible
				leukotactin-1	cytokine A15
				leukotactin 1	

Note: Information from the ImmuneXpresso project (http://www.immport-labs.org); with thanks to Shai Shen-Orr and Nophar Giefman.

These problems are compounded by the fact that clinicians, informaticians, and researchers do not always use terms systematically or consistently. To give just one example, the chemokine identified in the Protein Ontology as

PR:000001987 C-C motif chemokine 15

is referred to, in the literature of different research communities, using all of the labels listed in table 0.1.

Such incongruities cause problems not only for human experts—who are typically familiar with the usage only in their own disciplinary communities—but also for computers. Even small imprecisions of this sort, if multiplied throughout a database or information repository, can lead to problems in successfully finding, organizing, and using relevant information by means of computers. They can lead not only to faulty classifications of biomedical phenomena and to failures in communication of research results, but also to faulty diagnoses.

The Computer Limitations Problem
In addition, the very virtues of computer implementations—unlimited memory, efficient retrieval and reasoning, widespread availability—create a further issue, which we shall refer to as the *computer limitations problem*.

This problem arises from the fact that computers are, in familiar ways, unintelligent; they do not understand themselves, their programming, or the intended interpretation of the representations that they contain and manipulate. For this reason the use of computers in addressing the issues of scientific information management creates problems of its own. Above all, the very success of ontology-based approaches to the integration of data has led to a *multiplication of ontologies* in ways that threaten to recreate in a new form the very problems of interoperability that ontologies were themselves designed to solve.

Some Implications of Computer Limitations for Information Representation and Management

Computers are limited, first, in the sense that they cannot tell good from bad data. If data entered into a computer are described in vague or ambiguous or incoherent ways, then the computer programmed to reason with such data will likely produce results that are equally vague or ambiguous or incoherent. For example, one early version of the International Classification of Nursing Practice (ICNP) defined "water" as "a type of Nursing Phenomenon of Physical Environment with the specific characteristics: clear liquid compound of hydrogen and oxygen that is essential for most plant and animal life influencing life and development of human beings."[4]

A definition such as this, while clearly incorrect—water is not "a type of nursing phenomenon"—may seem relatively benign (after all, the definition is not completely incorrect). However, from the definition of "water" just provided, it would follow by relatively standard reasoning that "there is a nursing phenomenon that is essential for most plant life." To a human mind this is a puzzling and clearly false claim, but to an automated reasoning program it is a simple consequence of the definition. Such mistakes can severely compromise the goals of information aggregation, but they will not normally be detectable even by a perfectly functioning computer.

Second, two databases that store the same or related information, but use different terminologies or different organizational principles, cannot by themselves come to some kind of agreement about what they are referring to or carry out by themselves the task of aligning the information they contain. In such cases the data in the two repositories are siloed. Human beings, unlike computers, can comprehend the meaning as well as the vocabulary and grammar of a language. Thus, the information in two repositories such as those just described will be separated by a gap that can be crossed only with human intervention, for example through the provision of an explicit set of instructions for the translation (or "mapping") between the two terminologies. The National Institutes of Health refer in this connection to the phenomenon of "data wrangling," encompassing activities that make data more usable by changing their form but not their meaning: "Data wrangling may involve reformatting data, mapping data from one data model to another, and/or converting data into more consumable forms. Such data wrangling activities make it easier to submit data to a database or repository, load data into analysis software, publish to the Internet, compare datasets, or otherwise make data more accessible, usable, and shareable in different settings."[5]

All of these activities are designed to address what we might call the *Tower of Babel problem* of computers and information systems talking past each another in a mutual misinterpretation of information, which can arise even where computers use the same terms, if sameness of meaning has not been previously guaranteed.

What such problems show is that, while computers and information technology will indisputably form a central part of the solution to the problem of finding, organizing, integrating, and sharing the Big Scientific Data of the future, they are by no means the entire solution.

The Problem of Imprecise Thinking

The underlying goal, in all of the preceding, is the interoperability of data and information systems, or in other words a situation in which data collected in one system would be usable within the framework of a second system without further intervention by human beings. One specific obstacle to interoperability will be of primary importance in what follows—and is arguably the root cause of the other problems already discussed. It can be summarized under the heading of *imprecise thinking*, by which we mean a family of interrelated errors of logic and of language use that have repeatedly affected attempts to design information technology systems thus far. Avoiding these types of errors is important because, as discussed, it is ultimately human beings who are responsible for designing the ways in which data and information are imported into information systems.

Existing repositories of scientific data often contain not only ambiguities and inconsistencies in the use of technical terms, but also basic errors of logic. One reason for this is a matter of human resources; creating precise definitions of terms to be used in describing data, and ensuring consistent use of terms in accordance with these definitions, is expensive. It is also slow, and constrains the freedom to explore new kinds of data, a factor of considerable importance in areas of science that are advancing at a fast pace. In biomedicine and similar domains, therefore, we are caught in the horns of a dilemma. Each research community seeks to resolve issues of data representation in the most economical way, often by resorting to ad hoc shortcuts or workarounds, and the result is that basic errors of definition, of fact and of logic, creep into the systems that are developed in ways which degrade the quality and accessibility of the information they contain. This book is an attempt to help people break out of this dilemma.

An Example: The BRIDG Model

The following are examples of imprecise thinking taken from early versions of a database of biomedical information known as the Biomedical Research Integrated Domain Group (BRIDG) model. The BRIDG is a product of work by researchers from the National Cancer Institute (NCI), the U.S. Food and Drug Administration (FDA), the Clinical Data Interchange Standards Consortium (CDISC), and the Health Level 7 (HL7) Regulated Clinical Research Information Management Technical Committee (RCRIM TC) whose goal was to produce a "shared representation of the dynamic and static semantics" associated with protocol-driven research in the biomedical sciences and its associated regulatory artifacts. Part of BRIDG's work has included the creation of definitions for biomedical terms, many of which are taken over from its HL7 partner organization, including the definition of *living subject* as "a subtype of Entity representing an organism or complex animal, alive or not."[6]

The first problem with this definition is that it will cause logical problems when used in the coding of data about (for example) corpses, since it must leave room for coding data about an entity that is at one and the same time both living and not living.

The second problem is that it conflates an object (a living subject) with its representation (an "Entity representing"). This is like being told that a mammal is a *representation* of an animal that feeds its young with milk. It is an example of what philosophers refer to as the use/mention confusion, illustrated in its most blatant form by a sentence such as "Sleeping is healthy and contains three vowels." The word "mammal" is a noun that stands for or represents things, but mammals themselves are not nouns. Mammals are things, specifically living things that feed their young with milk.[7]

Another example is taken from an early version of BRIDG's definition of the class *performed activity* as "the description of applying, dispensing or giving agents or medications to subjects."[8] Here an activity, which is a kind of event that happens in the world, is defined as a "description" of certain other kinds of events. Here again there is an obvious confusion of an entity with the description of an entity.

Consider, now, what happens when the two definitions introduced so far are considered together for purposes of reasoning. Here are the definitions again:

• *Living subject:* a subtype of Entity representing an organism or complex animal, alive or not.

• *Performed activity:* the description of applying, dispensing, or giving agents or medications to subjects.

Suppose (as is plausible) that *administering insulin* is a *performed activity*, and that "subject" in the preceding refers to a *living subject*. Then, combining these two definitions, we get: *Administering insulin is the description of applying, dispensing, or giving agents or medications to a subtype of Entity representing an organism or complex animal, alive or not.*[9]

BRIDG, or at least its early version, is just one example of a terminology and modeling resource that has encoded such mistakes. In a more recent version *performed activity* is redefined as "an activity that is successfully or unsuccessfully completed."[10] Here the use-mention error has been corrected, but unfortunately it seems that a new error has entered in. To help the user, BRIDG provides two examples of usage of *performed activity*, as follows (where "CBC" stands for *complete blood count*):

CBC that is performed on a specific StudySubject on a given day.

A scheduled blood draw that is missed by a specific ExperimentalUnit on a given day.

We assume these to be examples of "successful" and "unsuccessful" completion, respectively. From the second example, however, it seems that we can infer that a blood draw that is *not performed* (because it is "missed") is an example of a *performed* activity.

Another common mistake in artifacts such as BRIDG is the inclusion of *circular definitions*. A circular definition is a definition that uses the term defined, or a close synonym, in the definition itself, thus rendering the definition (at best) uninformative. For example, the First Healthcare Interoperability Resources Specification (FHIR), again drawing on HL7, defined *container* as "a container of other entities" and *food* as "naturally occurring, processed, or manufactured entities that are primarily used as food for humans and animals."[11] Such circular definitions are not false (they amount to saying simply that a thing is what it is); but they are uninformative and therefore contrary to the intent of designing information resources that will be helpful repositories of information and will serve to constrain incorrect coding by providing information helpful to the coder. In some cases the definitions are worse than circular, for example when a term such as "health chart" is defined as "the role of a material (player) that is the physical health chart belonging to an organization (scoper)."[12] The best that can be said about definitions of this sort is that they certainly will not help ensure that the terms in question are used correctly by the human beings who need them in order to code healthcare data.

These are just a few examples of the kinds of imprecise thinking that can and do occur in biomedical informatics and other fields. Many of these errors in definition are familiar from subjects dealt with in logic and philosophy: problems of ambiguity, circular definitions, use-mention confusions, and confusions of reality with our thoughts or perceptions of reality. It is thus important to stress that there are principles of reasoning and methods—long familiar to students of logical philosophy—for constructing definitions and classifications of information in ways that are designed to avoid such errors. What is needed is for these principles to be articulated and brought to bear in a systematic way in the context of information systems.

Ontology as Part of the Solution

In philosophical contexts, "ontology" has traditionally been defined as the theory of what exists (or of "being *qua* being"): the study of the kinds of entities in reality and of the relationships that these entities bear to one another. This study includes in principle the entities dealt with by the special sciences (physics, chemistry, biology, and so on); but philosophical ontology has a universal focus. It is directed at the most basic or most general features of reality, features common to all domains, including, but not restricted to, the domains covered by science.

Examples of such general or common features of reality might include: unity and number, identity and difference, part and whole. The goal of philosophical ontology is to provide clear, coherent, and rigorously worked out accounts of such basic features, and to argue for these accounts, for example in light of their relative simplicity and logical coherence, and also in light of their coherence with the special sciences (or, in earlier times, for their coherence with theology).

In recent times use of the term "ontology" has become prominent in computer and information science, and ontology-based research has been especially successful in areas of bioinformatics. The term "ontology"—as in "Gene Ontology," "Infectious Disease Ontology," "Plant Ontology," and so forth—refers to a standardized representational framework providing a set of terms for the consistent description (or "annotation" or "tagging") of data and information across disciplinary and research community boundaries.

Ontologies are designed to promote greater consistency in description of data. To this end the terms in an ontology are provided with both textual definitions (to ensure consistency on the side of human beings maintaining and using the ontology) and logical definitions (to aid computer access and quality control). The result is organized in a graph-theoretic format (see chapter 4), in which terms serve as nodes in the graph and ontological relations (as between a type and its subtypes) serve as edges. The terms are then used for purposes of annotating or tagging heterogeneous data contained in multiple electronic information repositories.

In this way an ontology brings multiple advantages. It provides a common mode of access to the data. It promotes intertranslatability of the content of different data repositories, thereby promoting the cumulation of science. It helps to identify incompatibilities between different bodies of data, thus leading to new scientific questions, which may need to be addressed by new experiments. It enables the development of more powerful software tools for the mining of valuable scientific information from aggregated data stores and thereby also enables the formulation of more powerful queries and analytical methods.

For all of this to work, however, the ontologies themselves have to be developed in a mutually consistent fashion, they have to be used aggressively in annotations to multiple different sorts of data and information, and at the same time they have to be adjusted over time in such a way as to keep pace with scientific advance. The solution to the ontological problem is thus not simply a matter of agreeing on a common vocabulary and using it to annotate data. Rather, it requires a coordination of research groups simultaneously developing ontology resources in distinct but interrelated areas. Such coordination involves multiple different sorts of activity, including for example programming, project management, and user-interface development. Some of these activities, however, are of a philosophical nature. For experience has shown that the coordination of ontology developers working in different fields can be advanced if the ontologies they create and maintain through time can employ a common set of basic categories, which can be used as common starting point by developers of ontologies for different domains. This common set of basic categories helps in determining where entities of different kinds should be positioned within an ontology, to determine what relationships hold between them, and to determine how the corresponding terms in the ontology should be defined. The development and application of such a common set of basic categories is, however, itself a nontrivial exercise, and it requires the addressing of problems that are at least closely analogous to many of the traditional problems of philosophical ontology.

There is thus an affinity between the two senses of the term "ontology" we distinguished, and the recognition of this affinity allows us to define an approach to the building of computer-based ontologies on the basis of tried and tested logical and philosophical principles. This book is intended as an introduction to the foundations and methods of this approach.

A New Organon for the Information Age

In the following pages, we will present and explain the basic elements of the ontological approach to solving the problems of information management. In addition, we will put forward specific recommendations and principles that are intended for use by individuals interested in constructing ontologies in new domains.

While we believe that our proposals in this book are applicable in many other areas where terminologies and ontologies are being used in the organization of data and information, including industry, finance, government administration, manufacturing, and defense, we have directed our remarks primarily to those working in areas of information-driven scientific (including clinical) research. We view science as the systematic attempt to describe and explain what exists on the basis of experiments. Our primary focus will be on the construction of ontologies for the purpose of representing the sorts of entities that are of concern to scientific research. What is said can be applied, however, without restriction, to any domain where information is being assembled and used concerning what exists.

This document is thus conceived as an *organon*, by which is meant an *instrument*, for the conduct of scientific research, especially as concerns the representation of the results of such research in digital form. The first organon was written by Aristotle in the fourth century BC, and included his works on deductive reasoning (logic and the syllogism). Francis Bacon's *Novum Organum* (1620) was conceived as an extension and correction of Aristotle's *Organon* in light of the success of the experimental method introduced into science almost two thousand years later. Bacon focused on what is involved when inductive reasoning is used as part of a gradual approach to the understanding of nature, which in his eyes involves moving by degrees from particular cases and attempting to discover general axioms from those observations.

The increasing use of computers in scientific research, both for representing information and for acquiring novel results, has raised in its turn a host of new questions about the nature and methods of science. Our proposals in the chapters that follow are intended to constitute portions of a new instrument of scientific method for the information age, one that will clarify basic principles and methods of computer-based and computer-assisted scientific representation and research, with the goal of creating a more successful collaboration between scientists and informatics researchers.

We will focus on basic theoretical elements, on principles, and recommendations associated with best practices for the building of domain ontologies. Our subject, therefore, is the human contribution to the realization of the tasks involving ontology support in science, as opposed to issues associated with particular computational implementations. This is, first of all, because issues concerning idiosyncrasy and imprecision in human use of terms are logically prior to issues of computer encoding and implementation (the scientific information to be encoded must be properly understood, defined, and classified before it can be successfully encoded in a computer language). But it is also because the definitions, theory, and principles of best practice for ontology design that we will be discussing will in almost all cases be applicable regardless of the particular software framework in which the information is to be embedded. Our focus here is thus on giving researchers the theoretical principles they need to build useful ontologies that will be stable, we believe, even through successive generations of computer software and hardware.

Suggested Further Reading

For additional information on ontology and information ontology, see the website of Barry Smith, which includes ontology tutorials as well as links to both introductory and advanced essays on a range of ontological topics (http://ontology.buffalo.edu/smith/). We particularly recommend the following papers as introductions to basic issues of ontology and information ontology that reiterate and expand on the themes covered here in the introduction.

Feigenbaum, Lee, Ivan Herman, Tonya Hongsermeier, Eric Neumann, and Susie Stephens. "The Semantic Web in Action." *Scientific American* 297 (2007): 90–97.

Grenon, Pierre. "A Primer on Knowledge Management and Ontological Engineering." In *Applied Ontology: An Introduction*, ed. Katherine Munn and Barry Smith, 57–82. Frankfurt: Ontos Verlag, 2008.

Munn, Katherine. "Introduction: What Is Ontology For?" In *Applied Ontology: An Introduction*, ed. Katherine Munn and Barry Smith, 7–19. Frankfurt: Ontos Verlag, 2008.

Smith, Barry. "Ontology." In *Blackwell Guide to the Philosophy of Computing and Information*, ed. Luciano Floridi, 155–166. Oxford: Blackwell, 2003.

Smith, Barry, and Werner Ceusters. "Towards Industrial Strength Philosophy: How Analytical Ontology Can Help Medical Informatics." *Interdisciplinary Science Reviews* 28 (2003): 106–111.

Smith, Barry, and Bert Klagges. "Bioinformatics and Philosophy." In *Applied Ontology: An Introduction*, ed. Katherine Munn and Barry Smith, 21–38. Frankfurt: Ontos Verlag, 2008.

Smith, Barry, Waclaw Kusnierczyk, Daniel Schober, and Werner Ceusters. "Towards a Reference Terminology for Ontology Research and Development in the Biomedical Domain." In *Proceedings of the 2nd International Workshop on Formal Biomedical Knowledge Representation* (KR-MED 2006), vol. 222, ed. Olivier Bodenreider, 57–66. Baltimore, MD: KR-MED Publications, 2006. Accessed December 17, 2014. http://www.informatik.uni-trier.de/~ley/db/conf/krmed/krmed2006.html.

Smith, Barry, Lowell Vizenor, and Werner Ceusters. "Human Action in the Healthcare Domain: A Critical Analysis of HL7's Reference Information Model." In *Johanssonian Investigations: Essays in Honour of Ingvar Johansson on His Seventieth Birthday*, ed. Christer Svennerlind, Jan Almäng, and Rögnvaldur Ingthorsson, 554–573. Berlin/New York: de Gruyter, 2013.

1 What Is an Ontology?

Introduction

In order to design an ontology, it is important to understand just what an ontology is. Only on this basis can we be clear about both the steps that should be taken in ontology design and the kinds of pitfalls that should be avoided. The goal of this chapter and the next is to provide the basic definitions and distinctions in whose terms the process of ontology design can best be understood. Our definition of "ontology" is the following:

ontology = def. a representational artifact, comprising a taxonomy as proper part, whose representations are intended to designate some combination of universals, defined classes, and certain relations between them[1]

This definition employs a number of terms that are themselves in need of defining. Understanding these terms and the rationale behind their inclusion in the definition will take us a long way toward understanding what an ontology is.

The first term "taxonomy" we can define as follows (where here and in all that follows "universal" and "type" are used as synonyms):

taxonomy = def. a hierarchy consisting of terms denoting types (or universals or classes) linked by subtype relations

The most familiar kinds of taxonomies are those we find in biology (taxonomies of organisms into genera and species, as illustrated in figure 1.1). But taxonomies can be found also in any domain where it is possible to group things together into types or universals based on common features. We will discuss taxonomies in greater detail further on.

By "hierarchy" we mean a graph-theoretic structure (as in figure 1.1) consisting of nodes and edges with a single top-most node (the "root") connected to all other nodes

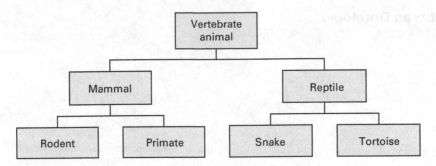

Figure 1.1
Fragment of a simple taxonomy of vertebrate animals

through unique branches (thus all nodes beneath the root have exactly one parent node).

By "types" or "universals" we mean the entities in the world referred to by the nodes (appearing here as boxes) in a hierarchy; in the case of figure 1.1, biological phyla, classes, and orders. As we will regularly use the term "entity" in a broad and generic sense, we here provisionally define it as follows:

entity = def. anything that exists, including objects, processes, and qualities

"Entity" thus comprehends also representations, models, images, beliefs, utterances, documents, observations, and so on.

Ontologies Are Representational Artifacts

Ontologies represent (or seek to represent) reality, and they do so in such a way that many different persons can understand the terms they contain and so learn about the entities in reality that these terms represent. Ontologies in the sense that is important to us here are designed to support the development, testing, and application of scientific theories, and so they will to a large degree be about the same sorts of entities as are represented by the general terms in scientific textbooks. Ontologies consist of terms arranged together in a certain way, and terms are an important subtype of representations:

representation = def. an entity (for example, a term, an idea, an image, a label, a description, an essay) that refers to some other entity or entities

When John remembers the Tower Bridge in London, then there is a representation in his mind that is about or refers to an entity other than itself, namely a certain bridge over the River Thames. Similarly, when Sally looks through a microscope at bacteria arrayed on a glass slide, then there are thoughts running through her mind to the effect that "these are *E. coli* that I am seeing." These thoughts involve representations that point beyond themselves and make reference to certain entities on the side of reality—in this case bacteria on the slide. It is one of the most basic features of human thought that beliefs, desires, and experiences in general point beyond themselves to certain entities that they are about. Note that a representation (for example, your memory of your grandmother) can be of or about a given entity even though it leaves out many aspects of its target. Note, too, that a representation may be vague or ambiguous, and it may rest on error.

Artifacts

artifact = def. something that is deliberately designed (or, in certain borderline cases, selected) by human beings to address a particular purpose

"Artifact" comes from the Latin *ars*, meaning "human skill" or "product." Artifacts include such things as knives, clothing, paperweights, automobiles, and hard drives. All artifacts are public entities in the sense that they can at least in principle be available to and used by multiple individuals in a community.

Representational Artifacts

Representational artifact = def. an artifact whose purpose is one of representation.

Thus a representational artifact is an artifact that has been designed and made to be *about* something (some portion of reality) and using some public form or format. Representational artifacts include things such as signs, books, diagrams, drawings, maps, and databases.

A key feature of representational artifacts of the sorts important to us here is that they come with rules for their interpretation. Maps do not come merely color coded, they also come with a legend or table that makes it possible to interpret their color coding as representing certain kinds of entities (countries, oceans, mountain ranges, etc.). Such legends have many of the features of ontologies, including the feature of supporting information integration; for example, maps that use a common legend can be more easily compared and combined.

A simple kind of representational artifact would be a drawing made by Sally of Tower Bridge based on her memory of how it looked when she visited London some years earlier. Sally's memory, and the images in her mind, are cognitive representations. Her drawing, in contrast, is a representational artifact that exists independently of such cognitive representations and transforms them into something that is publicly observable and inspectable. Just as Sally's memory of Tower Bridge can be better or worse, more or less accurate, so also the representational artifact that she creates on the basis of this memory can be better or worse, and more or less accurate as a representation of the entity to which it is intended to refer.

An ontology is an artifact, since it is something that has been deliberately produced or constructed by human beings to achieve certain purposes, and there is a sense in which—by analogy to Sally's drawing—it serves to make public mental representations on the part of its human creators. While not all representational artifacts are ontologies, all ontologies are representational artifacts, and thus everything that holds of representational artifacts in general holds also of ontologies.

Representational Units and Composite Representations
Representational units and composite representations are very common types of representations—encompassing practically the whole world of documents, which use written or printed language to represent things in the world. For example, the composite representation "John is drinking a glass of water," asserted by someone who is watching John, picks out a process in the world. The representational units in this composite representation include "John" and "glass"; these are the smallest referring bits of language contained within the sentence ("J," "w," and so on do not refer to or represent anything). Other examples of representational units include icons, names, simple word forms, or the sorts of alphanumeric identifiers we might find in patient records or automobile parts catalogs.

representational unit = def. a representation no proper part of which is a representational unit

composite representation = def. a representation built out of constituent subrepresentations as its parts, in the way in which paragraphs are built out of sentences and sentences out of words

Note that many images are not composite representations in the sense here defined, since they are not built out of smallest representational units in the way in which molecules are built out of atoms. (Pixels are not representational units since they are not representations.) Maps are typically built out of parts that include both representational units (for example, names of towns or hills) and image-like elements (for instance, shading used to represent inclines).

A Note on "Term"
In the following pages we will often make use of the word "term" to refer to the singular nouns and singular noun phrases that form the representational units and

composite representations in an ontology. The terms in an ontology are the linguistic expressions used in the ontology to represent the world, and drawn as nearly as possible from the standard terminologies used by human experts in the corresponding discipline. (Thus terms are distinct from identifiers of the sorts used in programming languages or from the alphanumeric IDs used in serial numbers or on credit cards.) Examples of terms in our sense include:

aorta	resident of Cincinnati	blood pressure
surgical procedure	smoking behavior	temperature
population	patient	blood glucose level

Terms in this sense can refer to single entities, collections of entities, or types of entities.

The question of what terms an ontology should include is determined (a) by the selected scope of the ontology (which is determined in turn by the purpose the ontology is designed to address), (b) by the available resources for population of the ontology, (c) by the structure of the domain that the ontology is intended to represent, and (d) by consensus among scientific experts about what the relevant entities are in that domain and about what they are to be called.

Ontology, Terminology, Conceptology

In our approach to ontology we assume that it is uncontroversial that ontologies should be understood as a kind of representational artifact, and that the entities represented are entities in reality—such as cells, molecules, organisms, planets, and so forth. Some ontologies contain terms which do not refer to any entities at all because—unknown to the developers—some type of error has been made. But even in those cases the terms in question are included in the ontology with the intention that they should refer. (Something like this was true, in former times, in the case of terms such as "phlogiston" and "ether.")

The relation between term and referent is to be understood by analogy with the relation of external directedness that is involved, for instance, when we assert that "Oxford" refers to Oxford, or "Ronald Reagan" refers to Ronald Reagan. This is true even where, as in ontologies such as the Mental Functioning Ontology (MFO),[2] terms refer to entities—for example, mental processes—that are internal to the mind or brain of human beings. Terms such as "mental process" too, as they appear in ontologies, are intended to refer to portions of reality in just the same sense as do terms referring to physical entities such as molecules or planets.

Confusion arises here in virtue of the fact that, in addition to the relation of reference or aboutness between terms in MFO and their mental targets in reality, there is another sort of relation between language and mind, which we can call the relation from term to concept. This latter relation holds in virtue of the fact that, when people use terms, they may associate these terms with mental representations—sometimes called "concepts"—of various sorts.

Ontology and Terminology: The Case of ISO

The relation from term to concept has played a central role in the discipline known as *terminology research*, which is in some ways a precursor to contemporary ontology. Terminology research grew as a means of coping with the large technical vocabularies used especially in areas such as commerce, manufacturing, and transport relevant to international trade, and the terminologist is interested in the usage of terms specifically from the point of view of standardization and of translation between technical languages. Terminology research is focused on concepts because, in the eyes of the terminologist, what is transmitted when a term is translated from one language into another is precisely some *concept*, which the users of the respective languages are held to share in common.

A view along these lines forms the foundation of the terminological work of the International Standards Organization (ISO), which pursues the goal of bringing about an "ordering of scientific-technical knowledge *at the level of concepts*" (emphasis added).[3] ISO hopes in this way to support the work of translators, and also to support the collection of data that is expressed in different languages. ISO Standard 1087–1, for example, sees terms as denotations of concepts, defining "concept" as follows:

A unit of thought constituted through abstraction on the basis of characteristics common to a set of objects.[4]

The background to this definition is a view of concept acquisition rooted in the phenomenalist ideas of the Vienna Circle.[5] Concepts are acquired, on this view, in virtue of the fact that, as we sense objects in our surroundings, we detect certain similarities—for instance between one horse and another, or between one red thing and another. We then learn to conceive the characteristics responsible for such similarities in abstraction from the objects that possess them.

Concepts are then formed through combination of such characteristics. Characteristics can be combined into concepts in many ways (for instance: {red, spherical}, {diseased, female, nonsmoker}, {with tomato sauce, with mozzarella, with pepperoni}), and for each such combination of characteristics there is, in principle at least, a corresponding concept. Equivalence between terms in different languages is a matter of correspondence between the corresponding bundles of characteristics. The terms are "equivalent," according to ISO, if and only if they denote one and the same concept.

What ISO leaves out of account—and what is left out of account by the ontologists who have been inspired by ISO—is the question of how we gain access to such concepts, entities that are alleged to exist at some language-independent level. Note, too, that ISO's own approach to standardization does not consistently follow an approach "on the level of concepts" of this sort. ISO Standard 3166–1, for example, defines a widely used set of codes for identifying countries and related entities. Currently ISO 3166–1 assigns official two-letter codes to 249 countries, dependent territories, and areas of geographical interest. The code assigned to France, for example, is ISO 3166–2:FR. And the code is assigned to France itself—to the country that is otherwise referred to as *Frankreich* or *Ranska*. It is not assigned to the *concept* of France (whatever that might be).

The Concept Orientation
We do not deny that mental representations have a role to play in the world of ontologies. When, for example, human biocurators use an ontology to tag data or literature or museum catalogs, then they will have certain thoughts or images in their minds. And if "concept" is used to refer to their understanding of the meanings of the terms they are using, then they can also be said to have concepts in their minds. Doctors, similarly, can be said to have concepts in their minds when diagnosing patients. Indeed, when a doctor *mis*diagnoses a patient then it is tempting to say that there was *only* the concept in his mind—and that there was nothing on the side of the patient to which this concept would correspond.

For this and other reasons, including the influence of ISO, the view of ontologies as representations of concepts has predominated especially in the field of medical or health informatics.[6] More recently, however, this "concept orientation" has been challenged by the "realist orientation" that is defended here.[7] The goal of ontology for the realist is not to describe the concepts in people's heads. Rather, ontology is an instrument of science, and the ontologist, like the scientist, is interested in terms or labels or codes—all of which are seen as linguistic entities—only insofar as they represent entities in reality. The goal of ontology is to describe and adequately represent those structures of reality that correspond to the general terms used by scientists.

Philosophical and Historical Background to Conceptualism
Another source for the view that the terms in ontologies represent our concepts of reality is epistemological, and draws on those strands in the history of contemporary ontology that connect ontology to the artificial intelligence/computer science field of

what is called "knowledge representation."[8] Since, it is argued, our knowledge is made up of concepts, representing knowledge—which means in this context roughly: representing logically the beliefs or the ontological commitments of scientists[9]—must imply representing concepts. This assumption in turn often goes hand in hand with a view to the effect that we cannot know reality directly or know the things in reality as they are in themselves, but rather that we have access to reality only as it is mediated by our own thoughts or concepts.

This is not a new view in the history of philosophical thinking about knowledge. Epistemological representationalism, for example, a view embraced by Kant, is the doctrine to the effect that our perceptions, thoughts, beliefs, and theories are most properly conceived as being about our constructions or projections, and only indirectly (if at all) about mind-independent entities in some external reality. Epistemological idealism, on the other hand, is a more extreme doctrine to the effect that our perceptions and thoughts are not about reality at all, but are entirely about mental objects such as perceptions, appearances, ideas, or concepts, because—for the idealist—that is all there is. In the formulation of the Irish philosopher George Berkeley, for example, "to be is to be perceived." Analogously, in the field of "knowledge-based systems," an ontology has been defined as "a theory of what entities could exist in the mind of a knowledgeable agent."[10]

Echoing such views, many in the field of knowledge representation have held that ontologies should be understood primarily as representing conceptual items. For example, Tom Gruber, the leader of the ontologist team that gave rise to the iPhone Siri app, influentially defined an ontology as "a formal specification of a shared conceptualization."[11]

Realism and Ontology

The view of ontology defended here, in contrast, is one according to which the terms in the ontology represent entities in the world—we might say that the ontology encapsulates the knowledge of the world that is associated with the general terms used by scientists in the corresponding domain.

There is a long and detailed history of debates in philosophy about whether we can have knowledge of an external world, and it is not our intent to rehearse these debates here. However, we can assert with confidence that representationalist and idealist positions are far from constituting the majority view among philosophers, either in the history of philosophy or in the philosophy of today. In regard to the latter some empirical evidence is provided by the results of the survey presented in table 1.1. Of the 931 philosophy faculty surveyed, only 4.3 percent supported idealism while 81.6 percent favored some form of nonskeptical realism.[12]

Table 1.1
PhilPapers survey results

External world: Idealism, skepticism, or nonskeptical realism?	
Accept or lean toward: nonskeptical realism	760/931 (81.6 percent)
Accept or lean toward: skepticism	45/931 (4.8 percent)
Accept or lean toward: idealism	40/931 (4.3 percent)
Other	86/931 (9.2 percent)

It is indeed true that we cannot perceive reality except by means of the specific sensory and cognitive faculties that we possess. But this in itself provides no reason for thinking that the experiences and concepts that we have do not provide us with information about reality itself. It would provide such a reason only if we had some evidence that our sensory and cognitive faculties were unable to apprehend reality—and this is precisely what is at issue.[13]

Certainly our cognitive faculties do not deliver the *entire* truth about reality; but this does not mean that the information that they do deliver should be viewed as nonrepresentative of how reality in fact is. For this, a separate argument is needed. On the position defended here, a version of *epistemological realism*, the most plausible way of understanding the relation between our cognitive faculties and reality is that our faculties—much like spectacles, microscopes, and telescopes—do indeed provide us with information about reality. They do this a little bit at a time, at different levels of granularity, and with occasional need for correction. One source of this correction is the application of the scientific method, which is itself an ongoing process of data collection and theorizing, using human perceptions supported by scientific experiments, and yielding results which are in their turn still fallible but also to a degree self-correcting in the course of time.

A further argument against the view of ontologies as representing concepts is heuristic in nature. It turns on the fact that acceptance of this view on the part of the developers of ontologies encourages certain kinds of errors, most prominently the sorts of use-mention mistakes that we have already touched upon in the introduction. The Systematized Nomenclature of Medicine (SNOMED),[14] a leading international clinical terminology, defined a "disorder" in releases up to 2010 as "a concept in which there is an explicit or implicit pathological process causing a state of disease which tends to exist for a significant length of time under ordinary circumstances." At the same time it defined "concepts" as "unique units of thought." From this it follows that a disorder is a unit of thought in which there is a pathological process causing a state of disease, so that to eradicate a disorder would involve eradicating a unit of thought. Recognizing its own confusion in this respect, versions of SNOMED since July 2010 have contained the warning:

Concept: An ambiguous term. Depending on the context, it may refer to the following:

- A clinical idea to which a unique ConceptId has been assigned.
- The ConceptId itself, which is the key of the Concepts Table (in this case it is less ambiguous to use the term "concept code").
- The real-world referent(s) of the ConceptId, that is, the class of entities in reality which the ConceptId represents.[15]

Accurately Representing Entities in Reality

What are the implications of our realist view for the understanding of representational artifacts such as ontologies and the terms they contain? Suppose, again, that Sally attempts to create a representational artifact that makes reference to Tower Bridge by drawing a picture. Our view is that it is here not the mental representation in her head, or the memories in her head, that Sally is trying to draw; rather, it is Tower Bridge itself. Should Sally have an opportunity to see the bridge again in the future and to compare it with the drawing that she has made, she may well identify a mistake or an absence of detail in the drawing and decide to correct it in order to create a more accurate representation—and this is so even if her original memory of the bridge contains no such additional information. Additionally, if other people look at the drawing of Tower Bridge and criticize its accuracy, they will engage in this criticism by citing facts, not about a memory or a mental representation, but about the drawing and about the bridge itself. Conceivably, Sally's memory may be in error, so that the drawing is discovered to be not of Tower Bridge but of, say, Chelsea Bridge. Then she would need, not to correct or enhance her drawing itself, but rather to assign it a new label.

All of this holds true, too, of the representations created by scientists. When constructing such a representation—whether it be a scientific theory presented in a textbook, or the content of a journal article or of a database—the goal is not to represent in a publicly accessible way the mental representations or concepts that exist in the scientists' minds. Rather, it is to represent the things in reality that these representations are representations of. When one queries the Gene Ontology Annotation Database, for example, in order to find out which HOX gene is responsible for antenna development in *Drosophila melanogaster*, then one is not interested in the conceptions or mental representations of the authors of the database or of the journal articles that lie behind it; rather, one is interested in the HOX gene itself, and in the process of antenna development in flies.

Respecting the Use-Mention Distinction

We have referred already to the use-mention distinction—the distinction between *using* a word to make reference to something in reality, and *mentioning* the same word in order to say something about this word itself. Thus it is one thing to consult (*use*) the periodic table in order to learn something about the chemical elements; it is quite another thing to talk about (*mention*) the periodic table as an important innovation in the history of human knowledge. We pointed out that confusion of use and mention is a common type of error in the building of ontologies—an error closely related to the view that terms in ontologies represent or denote concepts in people's minds.

All that is needed to avoid such errors is careful use of language. Thus one can use the phrase "Tower Bridge" to refer to an object in reality, as in "Tower Bridge is a well-known structure on the River Thames in London." However, one can also mention the same phrase, as in "'Tower Bridge' is used by speakers of English to refer to a structure on the River Thames in London" or "'Tower Bridge' is made up of eleven letter tokens of nine letter types from the Latin alphabet."

Similar considerations apply to the drawing of Tower Bridge discussed earlier. We can *use* such a drawing in order to explain to someone what Tower Bridge is, and what its characteristic features are. In this case, the drawing is being used as a representation of a certain bridge in London. But we can also *mention* the drawing, making it and its properties the explicit theme of discourse, for instance in "this drawing is made with paper and pencil," or "this drawing is 100 years old." We then make assertions that are about the representation itself, not about that to which it refers.

Use-mention errors are a very common mistake in terminology-focused areas of information technology—something that may have to do with the habit of many computer modelers to employ the very same terms for the elements of the models inside their computers as are used to refer to the real-world objects that these elements stand proxy for. As Daniel Dennett notes, computer and information scientists are often desensitized to use-mention problems because the objects to which their terms refer are entities that are properly at home inside the computer (or inside the realm of mathematical entities).[16] In this way refrigerators become identified with (are "modeled" as) refrigerator serial numbers; persons are identified with social security numbers. The following definition of "telephone" was proposed within the Health Level 7 (HL7) community in 2007: "Telephone: a telephone is an observation with a value having datatype 'Telecom.'"[17] As we shall see in chapter 4, a definition of a term in an ontology is a statement of the necessary and sufficient conditions an entity must satisfy to fall under this term. What one gets from HL7 and similar attempts to "model" healthcare reality is an explanation of how the term "telephone" could be used as a part of the representational artifact that is HL7. The use-mention conflation turns on the fact that a telephone is confused with the datatype of a certain representation of a telephone in a certain model. Microsoft HealthVault, similarly, defines a health record item as "a single piece of data that is accessible through the HealthVault Service";[18] it then defines "allergy" as a class that "represents a health record item type that encapsulates an allergy."[19] So, an allergy is defined not as a type of medical condition but rather as a piece of data within the Microsoft HealthVault.

Use-mention confusions need not be fatal in the hands of skilled computer modelers; they are, though, fatal when it comes to building coherent ontologies.

Ontologies Represent Universals, Defined Classes, and the Relations Between Them

At the beginning of this chapter we defined an ontology as "a representational artifact, comprising a taxonomy as proper part, whose representations are intended to designate some combination of universals, defined classes, and certain relations between them."

So far we have discussed in great detail the first part of this definition: the idea that an ontology is a representational artifact, and we have argued that the best way to understand the goal of building ontologies as representational artifacts is to see ontologies as representing entities in reality rather than concepts or other kinds of mental representations in the minds of human beings. We now turn to the second part of this definition, which specifies what it is in reality that is intended when we speak of "universals, defined classes, and certain relations between them."

The Goal of Science Is to Represent General Features of Reality

It is a basic assumption of scientific inquiry that nature is at least to some degree structured, ordered, and regular. Scientific experimentation involves in every case observations of particular instances of more *general types*—this *eukaryote cell* under that *microscope*, this *portion of H_2O* in that *flask*, the *cancer* in Frank's *body* (where the terms in italics pick out universals). "This" and "that" and "Frank," here, pick out instances that can be observed in the lab or clinic. The ultimate goal of science is to use observations and manipulations of such particulars[20] in order to construct, validate, or falsify general statements and laws; the latter will then in their turn assist in the explanation and prediction of further real-world phenomena at the level of instances.

Ontology is concerned with representing the results of science at the level of general theory (the generalizations and laws of science), not of particular facts. More precisely: it is directed at encoding certain sorts of information about the general features of things in reality, rather than information about particular individuals, times, or places.

Ontological Realism

The question thus arises as to what exactly is it that is general in reality, and what the general terms used by scientists in the formulation of their theories are supposed to be about. The question *what is it that is general in reality?* is roughly the question of what it is that makes scientific generalizations and law-like statements true. What do all the entities that a term such as "eukaryote cell" refers to have in common that makes them together form a type or universal?[21] Our preferred answer to this question, which we call ontological realism, says that there is some eukaryote cell *universal* of which all particular eukaryote cells are *instances*. On this view, universals are entities in reality that are responsible for the structure, order, and regularity—the similarities—that are to be found there. To talk of universals is to talk of what all members of a natural class or natural kind such as *cell*, or *organism*, or *lipid*, or *heart* have in common. Thus we capture the fact that the members of a particular kind are in some respect similar by asserting that they instantiate certain corresponding universals. Universals are repeatable, in the sense that they can be instantiated by more than one object and at more than one time, whereas particulars, such as this specific cell, your cat Tibbles, and the first city manager of Wichita, are nonrepeatable: they can exist only in one place at any given time.

In the history of Western philosophy, the form of realism that recognizes universals as existing in their instances in this way has roots in the work of Aristotle. According to Aristotle (on the simplified reading that we presuppose here), universals are mind-independent features of reality that exist only as instantiated in their respective instances. A mind can, by attention and abstraction, grasp universals instantiated by particular things—for example, the two universals *redness* and *ball* can be abstracted from the several particular red balls that we see lying around on the floor. These particulars have in common that they are all red and that they are all balls. The universals *redness*, *ball*, and *spherical shape* then exist in these particular instances that we see on the floor. For Aristotle, there must always be particulars (instances) that "ground" the existence of universals in the sense that the universals depend for their existence upon these particulars. On our view, it is such universals that are the primary objects of scientific inquiry, and thus also the primary objects to be represented in a scientific ontology.

We depart from Aristotle in a number of ways, however, many of which have to do with the fact that Aristotle lived before the Darwinian era. One important difference is that we allow universals not only in the realm of natural objects such as enzymes and chromosomes, but also in the realm of material artifacts such as flasks and syringes, and also in the realm of information artifacts such as currency notes and scientific publications.

Metaphysical Nominalism

The major alternative to realism about universals is nominalism. Nominalism is the claim that only particular (nonrepeatable) entities exist. There are no (repeatable) universals; nothing general or common from one object to another in reality at all. When we refer to some kind or category of things like cell, electron, molecule, or spherical shape, we are merely using a name (the term "nominalism" derives from the Latin *nomen* for "name") that stands for the plurality of relevant particular entities. There is, on this view, nothing in reality that is responsible for the order and regularity that we seem to observe in nature, or for the fact that things belonging to a kind exhibit similarity with respect to certain features or properties.

For extreme nominalists the general terms that we use pick out collections of entities in reality that are arbitrary in the sense that they reflect merely our chosen groupings of entities and the attachment thereto of general words or concepts. There are no bona fide joints or divisions in reality at the level of kinds; all judgments pertaining to what is general involve our having imposed some order on a reality that does not possess such order in and of itself.

As we have seen, there are some in informatics research who hold a view along these lines according to which an ontology is something like a conceptualization of reality—something that represents (say) a scientist's view of reality, rather than general features on the side of reality itself. At least some of the proponents of this view embrace it because they believe that we cannot know about reality but only about our own conceptualizations.[22] This leaves only concepts as objects of the general knowledge that the sciences aim to achieve (which seems to us to imply that the whole of science would be a branch of psychology, or linguistics).

The debate between realists and nominalists about the status of what is general in reality, too, has a long history, but we shall limit ourselves here to just two of the reasons why we adopt the realist side in this debate.

First, it is not at all clear that nominalists do indeed provide an account of how general terms and concepts can be applied to reality that does in fact avoid the appeal to things like universals. Consider, for example, the explanation of the biological category *mammal* proposed by the school of what are called "resemblance nominalists."[23] The word "mammal," they say, is a word that human beings have found it useful to employ in order to group together certain individuals in reality based on perceived or supposed similarities or resemblances among these things. Where an ontological realist will want to insist that there exists a genuine characteristic or feature (a universal) that is common to all of these things, the resemblance nominalist will insist that there exist only the individuals and the relations of similarity among them, and nothing more. But what is to be said about these "relations of similarity" or "resemblances" themselves? Presumably there is some "relation of similarity" R that obtains between all mammals on the one hand, and also some "relation of similarity" R* that obtains between all plants on the other. Clearly R and R* must be different relations, otherwise we would regularly mistake mammals for plants and conversely. So the question for the nominalist is: what is it that makes all instances of observed similarity R the same as one another and also different from all instances of observed similarity R*? While denying the existence of universals such as *mammal* or *plant* the nominalist is in danger of simply reintroducing them at the level of the different kinds of similarity relations holding among the corresponding different kinds of things.

A second point against nominalism is that it leaves us with no explanation of the success of science, which enables successful predictions precisely on the basis of general laws. We know of no way to understand this ability except by appeal to the assumption that science does this by concerning itself not with particular instances of things, but rather with repeatable features, forming general patterns or structures, that are instantiated in particular things. *Lipid*, for example, is a universal that scientists are able to identify not by virtue of what is specific to the fats in John's body, or the sterols in Professor Jones's lab, or the fat-soluble vitamins in the bottle at the local pharmacy, but rather by virtue of the universal features or characteristics shared by all of these particular instances.

Universals and Particulars

Particulars, in opposition to universals, are individual denizens of reality restricted to particular times and places. Particulars instantiate universals, but they cannot themselves be instantiated. In virtue of instantiating the same universal, two particulars will be similar in certain corresponding respects. Particulars exist in space and time. It is possible to interact with, directly see with one's eyes, as well as touch and smell,

Table 1.2
Borges's Celestial Emporium of Benevolent Knowledge

In his "The Analytical Language of John Wilkins," Jorge Luis Borges describes "a certain Chinese Encyclopedia," the Celestial Emporium of Benevolent Knowledge, in which it is written that animals are divided into

1. those that belong to the Emperor
2. embalmed ones
3. those that are trained
4. suckling pigs
5. mermaids
6. fabulous ones
7. stray dogs
8. those included in the present classification
9. those that tremble as if they were mad
10. innumerable ones
11. those drawn with a very fine camelhair brush
12. others
13. those that have just broken a flower vase
14. those that from a long way off look like flies

Source: Jorge Luis Borges, *Other Inquisitions: 1937–1952* (Austin: University of Texas Press, 2000), 101.

photograph, or weigh, particulars of many sorts. Universals, in contrast, are accessible only via cognitive processes of a more complex sort.

How do we establish whether a given general term (such as "H_2O molecule" or "cell" or "mammal" or "sport utility vehicle" or "former fan of ABBA") picks out a universal? The answer to this question is not an easy one to formulate (any more than would be the answer to questions such as "How do we establish whether a given statement is true?" or "How do we establish whether a given statement is something that is known to be true, or expresses a law of nature?"). However, just as we can distinguish clear cases of truths (that red is a color) and falsehoods (that the earth is shaped like a cube), so we can distinguish also certain clear cases of general terms that do designate universals (such as names of chemical elements) and certain clear cases of general terms that do not designate universals (such as the majority of the terms listed as designating types of animals in table 1.2).

And while we cannot give any algorithm for determining how such terms are to be identified, a number of rules of thumb for such determination are provided in box 1.1. If a single general term yields a positive answer to all, or almost all, of these questions then this is a strong positive indication that it refers to a universal.

Box 1.1
Does a General Term Pick Out a Universal?

1. Can we point to (or identify through experimental assays) entities in the realm of particulars to which the term applies (entities which would then be instances of the universal in question)?
2. Is the referent of the term repeatable, or multiply instantiable, in the strong sense that it can have an open-ended family of instances?
3. Is the general term in question logically nondecomposable into simpler general terms, as contrasted, for example, with a general term such as "either female or carrying an umbrella"?
4. Is the general term in question, or some synonym thereof, used in the formulation of multiple scientific laws?
5. If the term is composed of other terms, are all of these themselves general terms? If yes, do these general terms (or their ultimate general-term constituents) yield positive answers to the other questions on this list?

The decision as to whether a given term does designate a universal may, however, in every case be revised. We may discover, for instance, that a general term refers to multiple distinct diseases (as was the case with "diabetes" and "hepatitis"). Such revisability is not, however, a concern relating specifically to our treatment of universals or of the general terms in ontologies; rather, it is an ineluctable feature of science as a whole.

Empty or Potentially Empty General Terms
An ontology is a representational artifact whose purpose is to represent what is general in reality. An ontology, in other words, is concerned with representing universals. At any given stage in its development science will, for many general terms, give us confidence in the belief that the terms in question designate universals. On the other side, there are some candidate ontology terms where we have similar confidence that they do not denote a universal—for example, "unicorn" or "perpetual motion machine" or "regular smoker and is identical to some prime number." Such terms do not designate anything at all in the strong sense that there are no particulars to which the terms in question can be correctly applied.

There are, however, examples of general terms for which it is not clear whether or not they designate universals. In the eighteenth century this was briefly the case with the term "phlogiston," until the term fell out of favor; until recently it was the case with the term "Higgs boson." Such cases arise particularly in those areas where at any given time the most exciting scientific advances are taking place. In general, ontologies

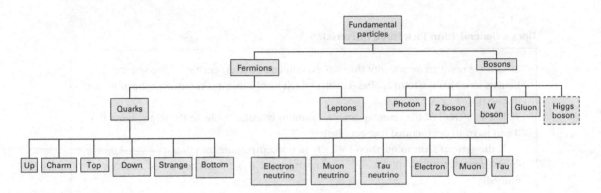

Figure 1.2
An experimental ontology created when use of the term "Higgs boson" was still considered speculative

will be created to capture the content of established scientific theories—the sort of content that is expressed in textbooks for use by new generations of scientists who want to learn the general theoretical framework that forms the basis of new and controversial hypotheses and methods. Ontologies can in such cases be created experimentally, in order to capture the content of one or more of the alternative hypotheses currently being explored at the fringes of established science. But use of a term in such an ontology remains tentative—in the sense that no ontological commitment is involved—until scientific disputes are resolved and either the term falls out of favor or some referent is securely attached.

Such experimental, or provisional, ontologies (see, for example, figure 1.2) are the equivalent of setting aside terms or codes for future use, for example, when creating a database of serial numbers for items in production. Some serial numbers will be used, in due course, for tracking items actually produced; others might be used for inventory planning or similar purposes, but again in a way that remains tentative—in the sense that no ontological commitment is involved.[24]

Universal vs. Class
An additional set of problems is created by those general terms often used by science to refer to particulars in reality but in relation to which (in light of our questions in box 1.1) there is no corresponding universal. Examples are "smoker in Leipzig," "person of the Hindu religion who has bathed in the Ganges," "Finnish spy," and so on.

To see how, within the realist framework, such cases are to be dealt with we start with the distinction between universals on the one hand, and the classes which form their extensions on the other. The universal *cell membrane*, for example, has as its extension the class of all cell membranes. It is not only universals that have extensions, but also general terms. The extension of the general term "cell membrane" is identical to the extension of the universal *cell membrane*; however, even those general terms for which there is no corresponding universal providing they are nonempty will have extensions also.

A class, on our view, is defined as a maximal collection of particulars falling under a given general term. All extensions of (nonempty) general terms are classes. (We leave open the issue as to whether empty general terms have extensions.) Thus the class of mammals, for instance, is the maximal collection of all mammals. The class of H_2O molecules is the maximal collection of all H_2O molecules. The term "mammal" applies to every member of this class, and every particular to which the term applies is a member of this class. Each universal has a corresponding maximal class as its extension. We might call such classes "natural classes." The class of all human individuals with less than an inch of hair on their heads picks out a class of individuals in reality, but it is not a natural class, and so there is little reason to think of this class as corresponding to a universal.

But there may be reasons nonetheless to include such a term in an ontology. In performing clinical research, for example, we may have data pertaining to "human beings diagnosed with hypertension," "human beings born in Vermont," "human beings whose mother has died," and so on. Classes corresponding to terms such as these are demarcated on the basis of selection criteria defined by human beings. Thus we will refer to them in what follows as "defined classes."

There are at least two recognizably distinct families of defined classes:

1. *Classes defined by general terms abbreviating logical combinations of terms denoting universals.* These classes can be divided into two groups:

a. defined by selection: for example, *woman with green eyes, protein molecule which has undergone a process of phosphorylation, disinfected scalpel*. Such classes are subclasses of extensions of given universals, and in the simplest cases they are defined through logical conjunction; they often involve features pertaining to the histories of the entities in question, and include cases where what transpired historically leaves no physical change in the entities in question;

b. defined by combination: here the class definition comprehends members which instantiate two or more nonoverlapping universals, for example, *current cost item* (defined as either *cash* or *account receivable*), *employee* (defined as either *waged employee* or *salaried employee*). Such classes are unions of extensions of given universals, and in the simplest cases they are defined through logical disjunction;

2. Classes defined by general terms abbreviating logical combinations of terms denoting universals with terms denoting particulars, for example, *woman currently living on the north coast of Germany, male athlete born after 1980, individual in the Western Hemisphere currently infected by HIV.*

Note that in some cases of terms involving logical combinations of terms designating universals, the resulting compound terms will themselves designate universals. The terminological practices of biology, which are our best guide as to what universals biologists are committed to ontologically, tell us that the term defined through the conjunction of the two universals *eukaryote* and *cell* itself refers to a more specific universal *eukaryote cell.* These terminological practices tell us also, however, that even though *mammal* and *electron* are universals, there is no universal *mammalian electron*—the term in question does not track scientifically interesting similarities among the entities in nature. (Thus also it would be a mistake to include this term in an ontology that is designed to support scientific reasoning.) The term "mammalian electron" might, however, be added to an ontology to support a particular application. It would then refer to a defined class falling under subfamily (1a) in our list.

Similarly, terms defined as described in (1b) pick out defined classes rather than universals because of their explicit inclusion of a disjunction as a primary feature of their definitions. A term such as "mammal or bacterium" does not designate a universal because such a term does not pick out a scientifically interesting collection of entities. Disjunctive terms of this sort, while again they may be useful for certain practical purposes, will in almost all cases refer only to what we can think of as gerrymandered classes.

To see why terms defined as described in (2) do not pick out universals, consider the expression "woman currently living on the north coast of Germany." This refers to a particular collection of particular women in a specific location and at a specific time. Working through the criteria provided in box 1.1: it is possible to point to individuals who are instances (or better: members) of this class (1), however the characteristics identified in this class are not open-endedly repeatable (2), do not contain only general terms (3) ("currently" and "north coast of Germany" refer to particular times, places, and countries), and surely do not figure prominently in any scientific laws or theories (4). Again, such classes are often of interest to scientists working on particular issues or problems, for example in the context of public health analysis or clinical trials involving specific subject populations. A scientist may, for example, be studying incidence of diabetes in nonsmoking juveniles in downtown Baltimore born in a given year. However, any scientific conclusions drawn on the basis of such trials will be general in nature, and will be formulated by reference to universals in the sense we have outlined in the preceding.

Universals, their extensions, and defined classes are all important for ontological purposes. However, as will be made clearer in the chapters to follow, it is essential that they be carefully distinguished, and that primary importance should be given in ontology construction for purposes of scientific research to the accurate representation of universals.

Relations in Ontologies

The final element in our definition of an ontology is the reference to the relations holding among universals and defined classes. The general idea of a relation is familiar from common sense. A woman typing on a laptop in a Manhattan coffee shop stands in several relations to several other entities, and each one of those other entities is involved in multiple relations to further entities. She is

- an **instance of** organism,
- the **daughter of** a stockbroker,
- such as to **exemplify** the quality of being seated,
- **supported by** a chair,
- **adjacent to** the counter,
- **located in** Manhattan,
- **married to** her spouse.

At the same time,

- her arm is **part of** her body,
- her laptop screen is **part of** her laptop,
- her latte **is colder than** that of her neighbor,

and so on, in ever-widening circles.

Similarly, if a bridge collapses immediately after an explosion directly underneath it, then we can assert that the explosion event stands to the bridge-collapse event in the relation of cause to effect, that the explosion event occurred at a certain time, that the bridge-collapse event occurred at a certain later time, and so on.

In building ontologies, however, we are interested not only in relations of these sorts that hold among instances, but also in relations that hold between the corresponding universals, as for example between types of organisms and their typical anatomical parts, between types of events and their temporal locations, between types of events and the types of objects that participate in them, and so forth. For example, it is universally the case that every instance of mammal *includes as part* some instance of brain.

Assertions about such relations form a major part of scientific knowledge. It is one thing to know something about the genus *feline*; it is a much better thing to know also how the genus feline fits into the larger picture of living things in nature—in particular, what its relation is to other genera, to the associated genes, cells, organs, and habitats. Similarly, it is one thing to understand something about the universal *hydrogen*; but it is another thing to know how hydrogen is related to other elements, to the types of molecules of which it forms a part, to the behaviors of such molecules in given types of reactions, and so on.

The representation of universals in ontologies involves representation also of the relations in which these universals stand to other universals, and this fact differentiates them from *terminologies*, conceived as representational artifacts containing lists of lexical entries and descriptions thereof, but which do not render formally explicit the relations holding among the entities referred to by these entries.

Basic Relations

We will deal with relations at length in chapter 7, but for the moment it is useful to distinguish three different kinds of binary relations, which will play an important role in the discussions that follow:

- relations that hold between two universals;
- relations that hold between a universal and a particular;
- relations that hold between two particulars.

Universal-Universal Relations

1. The paradigm example of a relation that holds between universals is the *is_a* (meaning "is a subtype of") relation, as in

protein molecule *is_a* molecule,

explosion event *is_a* event,

and so on.

The *is_a* relation holds among universals in virtue of the fact that universals stand in hierarchies of generality (referred to as "taxonomies," below). For example, the hierarchy extending from the universal *tiger* (for example, *panthera tigris tigris*) through the universals *panthera, feliformia, mammalia, chordata*, and finally to *cellular organism, living thing*, and *object* can be understood as structured from least to most general in terms of the *is_a* relation, as in table 1.3. Thus, more specific ("child") universals stand in *is_a* relation to more general ("parent") universals.[25]

Table 1.3
Examples of the *is_a* relation with a taxonomic hierarchy

- *panthera tigris tigris is_a panthera feliformia*
- *panthera feliformia is_a mammalia*
- *mammalia is_a chordata*
- *chordata is_a cellular organism*
- *cellular organism is_an object.*

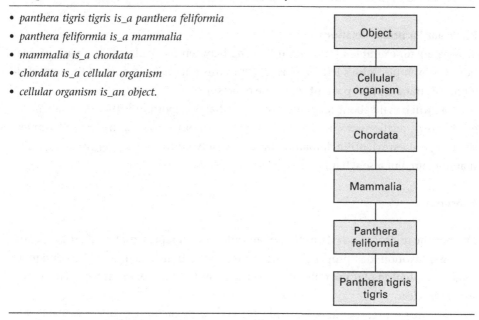

Universal-Particular Relations

2. A paradigm example of a relation between a particular and a universal is the **instantiates** relation, as in Barack Obama **instantiates** *human being*, where Barack Obama is the particular flesh and blood entity living in the White House, and *human being* is the universal. Other examples include

- these particular stellar cells under the microscope **instantiate** the universal *stellar cell*
- that oak tree on the corner of Main and Elm **instantiates** the universal *oak tree*

All particulars stand in the instantiation relation to some universal—in fact, typically to several universals at different levels of generality—but universals themselves do not instantiate anything. However, no particular stands in an *is_a* relation to any entity. Further examples of relations holding between a particular and a universal include **is allergic to** (as in John **is allergic to** *penicillin*), **knows about, is an expert on** (as in Mary **is an expert on** Lepidoptera), and other relations involving mental directedness.

Note that we are following here and in what follows the convention that for relational assertions involving one or more particulars the corresponding relation is picked out in **bold**; for assertions involving only universals we use *italics*. (These conventions are explained in more detail in chapter 7.)

Particular-Particular Relations

3. A paradigm example of a relation holding between particulars is the **part_of** relation. For example, John's left leg **part_of** John, this microtubule **part_of** that cytoskeleton, this transcription **part_of** that gene expression.

More will be said about relations. For now, what is important is that there are different kinds of relations, some of which hold among universals, and that fully understanding a given scientific domain requires not only knowing what universals exist in that domain, but also what kinds of relations hold between them.

Conclusion

By now, the import of our definition of an ontology—a representational artifact, comprising a taxonomy as proper part, whose representations are intended to designate some combination of universals, defined classes, and certain relations between them—should be clear.

Ontologies are representational artifacts in the sense of being publicly available representations of scientific information about reality. In their role as representations of science, their primary purpose is to represent general features of reality, what we have called universals, and the relations that exist between them. In addition, because defined classes are often useful in science, these too will often be represented in ontologies along with the relations obtaining among them. As we shall see in chapter 8, much current ontology work deals with ontologies as artifacts formulated using the Web Ontology Language (OWL), which allows universals and defined classes to be treated identically as "classes" in the technical sense embraced by OWL. This does not imply, however, that the special role of universals emphasized in the preceding is of no consequence for OWL ontologies. For—as we shall be arguing throughout this work—to build an ontology that is able to serve the purposes of scientific research, it is vital that the ontology is built in such a way as to represent as accurately as possible the universals in the corresponding domains of reality. This holds even when we build ontologies incorporating terms representing defined classes. For in such cases, too, the terms representing the universals used in the definitions of these classes must be included, either in the relevant ontology, or in some neighboring ontology with which it is interoperable. Only in this way can we provide both the authors and the users of the ontology with a coherent view of what its terms refer to. In chapter 2, we will discuss in this light both the different kinds of ontologies, and also introduce the notion of a taxonomy and the crucial role that taxonomies play in the structuring of ontologies.

Further Reading on Issues of Epistemological and Ontological Realism

Armstrong, David. *Universals: An Opinionated Introduction*. Boulder, CO: Westview Press, 1989.

Johansson, Ingvar. *Ontological Investigations: An Enquiry into the Categories of Nature, Man, and Society*. New York: Routledge, 1989.

Lowe, E. J. *A Survey of Metaphysics*. Oxford: Oxford University Press, 2002.

Lowe, E. J. *The Four Category Ontology: A Metaphysical Foundation for Natural Science*. Oxford: Oxford University Press, 2006.

Smith, Barry. "Beyond Concepts: Ontology as Reality Representation." In *Formal Ontology in Information Systems: Proceedings of the Fourth International Conference (FOIS 2004)*, ed. Achille C. Varzi and Laure Vieu, 31–42. Amsterdam: IOS Press, 2004.

2 Kinds of Ontologies and the Role of Taxonomies

An ontology, as we conceive it, is a representational artifact aimed at representing universals, defined classes, and relations among them. We are interested here specifically in the universals, defined classes, and relations discovered by and pertinent to scientific research. In this chapter we discuss in more detail the philosophical background of ontology in general, and introduce some distinctions between kinds of ontologies, including domain ontologies, top-level ontologies, reference ontologies, and application ontologies. We also discuss in more detail the idea of a structured classification or taxonomy.

Philosophical Ontology

Historically, ontology is a branch of philosophy having its origins in ancient Greece in the work of philosophers such as Parmenides, Heraclitus, Plato, and Aristotle. "Ontology" derives from the Greek words *ontos* (meaning "existence" or "being") and *logos* (meaning "rational account" or "knowledge"). Ontology in this sense is concerned with the study of what is, of the kinds and structures of objects, properties, events, processes, and relations in reality. From the philosophical perspective, ontology seeks to provide a definitive and exhaustive classification of entities in all spheres of being. Philosophers have emphasized in this connection the role of certain basic or preferred types of entities—for example, absolute simples—that are regarded as being truly real, as contrasted with other, less-privileged entities, which are seen as being constructed on their basis.

Contemporary philosophical ontology (sometimes called "analytic metaphysics") has sometimes allowed the study of entities dealt with by the special sciences (physics, chemistry, biology, psychology, and others), but it still overwhelmingly pursues a more general focus, one directed at providing a description and explanation of the kinds of objects and relations that are common to all scientific domains. Examples of such common or domain-neutral features of reality include unity and plurality; cause and effect; identity, both at a time and across time; compositional structure, determined either via part-whole or member-set relations; spatial and temporal and spatiotemporal location; and so on.

Where the biologist studies cells, the chemist studies molecules, and the physicist studies energy and electrons; the philosophical ontologist, in contrast, is interested in giving an account of what is common to cells, molecules, and electrons (for example, that they are all entities or things that exist as the bearers of certain properties or qualities) and of the relations in which these kinds of entities stand to one another. Such relations may extend across the usual disciplinary boundaries of the special sciences and across granularities, as in the case of relationships between microlevel entities such as atoms or molecules and macrolevel entities such as organisms and planets. The goal of philosophical ontology is at its core to provide a clear, coherent, and rigorously worked-out account of the basic structures of the whole of reality thus conceived.

Philosophical Ontology and Taxonomy

A major feature of philosophical ontology is the classification of different kinds of entities. This involves not only differentiating between basic categories of entities (for example, between *objects* such as a heart, and *processes* such as the beating of a heart), but also recognizing more specific kinds of entities falling under these categories (such as the difference between an individual object, such as a human being, and an aggregate or group of objects, such as a population of human beings). The result is a set of representations of the kinds of objects there are, either in general or in a given field of investigation, organized according to certain principles of classification. The so-called Porphyrian Tree (illustrated in figure 2.1), a version of which was included in an introduction to the work of Aristotle by the Greek philosopher Porphyry, is designed to represent the principle types of things (in Greek the *katēgoriai*) found in reality from the Aristotelian point of view. This idea has been used by thinkers in a variety of disciplines ever since. The Linnaean taxonomy—first developed by Carl Linnaeus in the eighteenth century to classify living things and still used by biologists today—is based on the Porphyrian Tree, and the idea has been reused in multiple further fields—in the periodic table of the elements, the WHO International Classification of Diseases,[1] the Department of Defense Joint Hierarchy of Military Doctrine,[2] and many more.

Simple Taxonomies

The Porphyrian Tree is an example of what we have already referred to in chapter 1 as a *hierarchy*: a graph with *nodes* (including *leaves*, the lowest nodes) and *edges* (the lines

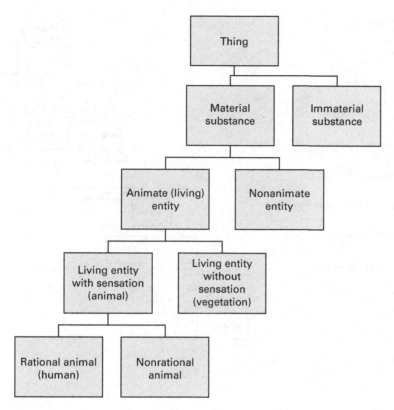

Figure 2.1
The Porphyrian Tree

connecting the nodes), forming *branches* that connect each node (each box in figure 2.1: thing, material substance, etc.) in a path leading upward to the highest node or *root* (see chapter 4 for further details). Nodes are connected by relations of subsumption (where the more general, or parent node, subsumes the less general, or child node).

All the ontologies with which we have to deal in what follows can be viewed as graph-theoretic structures along these lines, consisting of nodes (the terms of the ontology) connected by edges (representing relations). In a taxonomy the nodes are intended to represent types or universals in reality, while the edges represent *is_a* (which means subtype) relations connecting these types or universals in reality. We use *"is_a"* to mean "is a subtype of," defined as follows:[3]

A is_a B = def. *A* and *B* are types and all instances of type *A* are also instances of type *B*

Homo sapiens (species) is_a Homo (genus),

Homo is_a hominid (family),

hominid (family) is_a primate (order),

primate (order) is_a placental (subclass),

placental (subclass) is_a mammal (class),

mammal (class) is_a vertebrate (subphylum),

vertebrate (subphylum) is_a chordate (phylum),

chordate (phylum) is_a animal (kingdom),

animal (kingdom) is_a eukaryote (domain).

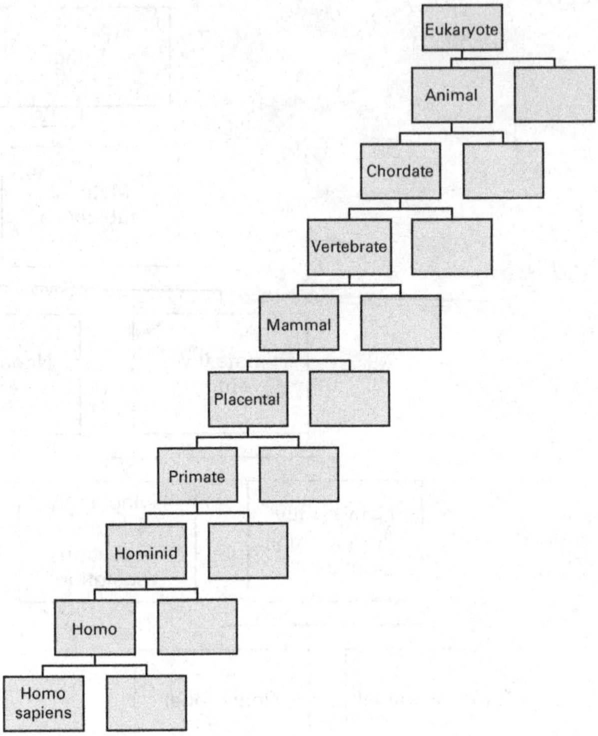

Figure 2.2
A partial Linnaean taxonomy

The basic *is_a* relations represented by the links connecting lower and higher nodes together form a hierarchical classification of entities. For example, in the Porphyrian Tree we have the following:

material substance is_a thing,

animate (living) entity is_a material substance,

living entity with sensation is_a living entity,

rational animal is_a living entity with sensation,

human is_a rational animal.

We have something similar in the Linnaean taxonomy shown in figure 2.2.

Both the Porphyrian and Linnaean taxonomies are built on the basis of an Aristotelian approach, which in effect conceives the *is_a* relation as a generalization of the species-genus relationship. In the Porphyrian Tree, *material substance* can be considered both as a species of the genus *thing*, and as a genus with *animate entity* and *nonanimate entity* as its two species. *Animate entity*, similarly, is the genus for both *living entity with sensation* (a species of animate entity) and *living entity without sensation* (another species of animate entity), and so on, down the tree to *human*. Importantly, at each step from a genus to a lower species, the species must be identified in terms of its possession of a *differentia*, some defining characteristic or characteristics that makes the species both more specific than its subsuming genus and serves to set it apart from other species of the same genus. In Porphyry's taxonomy being a living thing is what makes *animate entity* more specific than *material substance* and also what differentiates it from *nonanimate entity*. Similarly, being rational is what makes *rational animal* more specific than *living entity with sensation* and differentiates it from other *nonrational* types of animals. (As we will discuss in more detail in chapter 4, stating the genus and differentia of a type is a crucial part of providing what we shall refer to as an "Aristotelian definition" of that type. Taxonomies and definitions are closely interconnected in ontology design.)

Classification of the Porphyrian sort is nothing unusual. We are dealing constantly with hierarchical relationships of greater and lesser generality, whether in library catalogs, in restaurant menus, in functional classifications of genes, or in the directory structure used by the operating systems of our computers. Such sorting is a key element in the creation of an ontology, and constructing coherent taxonomies based on explicit and consistent principles of classification —set forth in detail in chapter 4—is one key component of good ontology design.

Formal vs. Material Ontologies

A formal ontology is domain neutral. It contains just those most general terms—such as "object" and "process"—which apply in all scientific disciplines whatsoever. Thus it corresponds to the sort of ontological interest we identified above as predominating among philosophers. A material (or "domain") ontology is domain specific. It contains terms—such as "cell," or "carburetor"—which apply only in a subset of disciplines.

A formal ontology is a representation of the *categories* of entities and of the relationships within and between them. We here adapt Aristotle's term "category" and use it to mean: domain-neutral universal. Categories are those universals whose instances are to be found in any domain of reality. Thus categories are also very general universals. The most obvious category is *entity*, meaning: anything that exists in any way. No matter what science one is considering it studies entities, and thus the category *entity* applies to the subject matter of that science. An example of a formal-ontological relation is *part_of,* since every science is committed to the existence of at least some entities which have or are parts.

A material ontology, by contrast, consists of representations of the material (meaning: nonformal) universals that are instantiated in some specific domain of reality, such as genetics, anatomy, plant biology, cancer, and so on. Where a formal ontology will contain representations of universals shared by many ontologies, each material ontology will contain representations used in the ontology alone. Thus the universals represented in an ontology of cell biology, for example, will not overlap with those represented in an ontology of cosmology or of architecture. This distinction will prove to be of great significance for the project of ontology design in support of information-driven science. Material or domain-specific ontology is discussed more in the following section.

Domain Ontology

A *domain* is a delineated portion of reality corresponding to a scientific discipline such as cell biology or electron microscopy, or to an area of knowledge or interest such as the Great War or stamp collecting or construction permits. Not everything that is part of an entity within a given domain is also part of that domain. Thus every human being has molecules as parts, but molecules do not form part of the domain of, for example, human geography, or human rights law (an issue that will be treated under the heading of "granularity" in chapter 3). Each domain ontology consists of a *taxonomy* (a hierarchy structured by the *is_a* relation) together with other relations such as *part_of, contained_in, adjacent_to, has_agent, preceded_by,* and so forth, along with definitions and axioms governing how its terms and relations are to be understood. A domain ontology is thus a taxonomy that has been enhanced to include more information about the universals, classes, and relations that it represents. A domain ontology provides a controlled, structured representation of the entities within the relevant domain, one that can be used, for example, to annotate data pertaining to entities in that domain in order to make the data more easily accessible and shareable by human beings and processable by computers.

Figure 2.3 represents a small section of a domain ontology for lipids (note its explicitly taxonomic structure).[4] A lipid is defined in this ontology as a hydrophobic or amphipathic small molecule, and this is consistent with the classification shown in the figure, where we can see that *"lipid is_a small molecule."* Notice that these researchers have not only chosen to classify various subtypes of lipids—for example, *LC lipoxins, LC hepoxilins,* and *LC cluvalones*—but have chosen to do this within a more general schema that is rooted in the class *biological entity*.

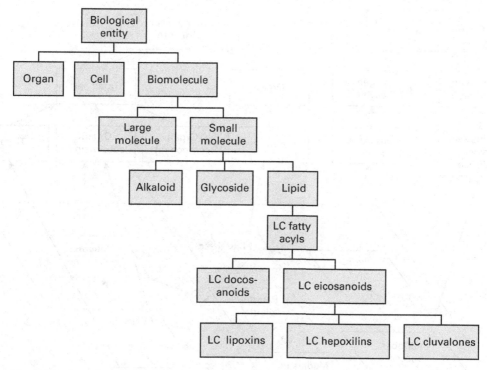

Figure 2.3
Section of a taxonomic classification from a lipid ontology

Domain Ontology and Taxonomy

The basic role of taxonomic classification in ontology has already been introduced. Here we discuss this issue in more detail as it relates specifically to domain ontologies. We have seen examples of taxonomies already. What we have called a *taxonomy* is a representational artifact that is organized hierarchically with nodes representing universals or classes and edges which represent the *is_a* or subtype relation. Where simple taxonomies are organized in terms of the basic *is_a* relation only, *ontologies* are organized also by other relations, such as *parthood*. For example, a domain ontology for cell biology might include information such as

nucleus is_a intracellular membrane-bounded organelle,

Golgi apparatus part_of plasma cell,

hindbrain nucleus part_of hindbrain

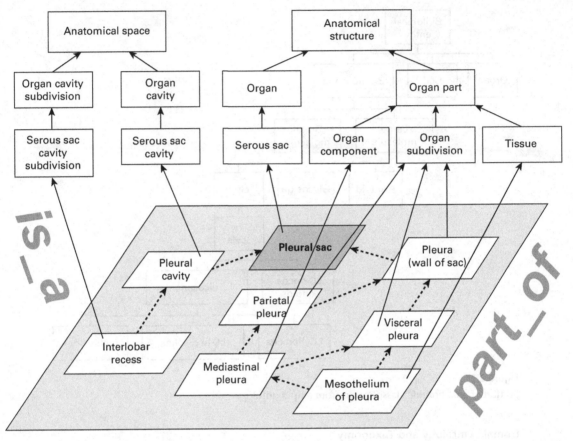

Figure 2.4
Fragment from the Foundational Model of Anatomy ontology
Source: C. Rosse and J. L. V. Mejino Jr., "The Foundational Model of Anatomy Ontology," in *Anatomy Ontologies for Bioinformatics: Principles and Practice*, vol. 6, ed. Albert Burger, Duncan Davidson, and Richard Baldock (London: Springer, 2007), 59–117.

and so on. This feature of ontologies is illustrated in figure 2.4 for the case of the Foundational Model of Anatomy. The solid arrows pointing vertically in the diagram represent the *is_a* relationship, while the dotted arrows moving horizontally across the bottom of the diagram represent the *part_of* relationship.

Similarly, figure 2.5 presents the beginnings of an ontology in the domain of biomedical ethics. What both of these ontologies have in common is that they explicitly represent multiple different kinds of relationships (in addition to species-genus hierarchies organized by the *is_a* relation) obtaining among the universals and classes that

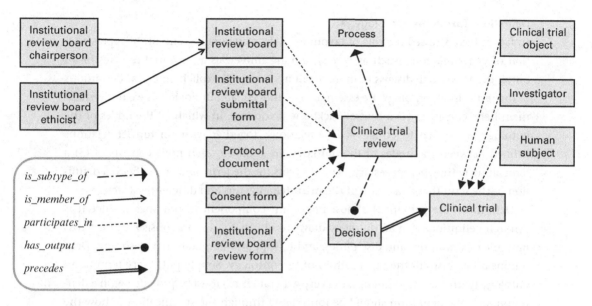

Figure 2.5

Section of a domain ontology for clinical trials in biomedical ethics

Source: David Koepsell, Robert Arp, Jennifer Fostel, and Barry Smith, "Creating a Controlled Vocabulary for the Ethics of Human Research: Towards a Biomedical Ethics Ontology," *The Journal of
Empirical Research on Human Research Ethics* 4 (2009): 43–58.

they represent. These additional relationships are indicated in the legend inside the
rounded rectangle in figure 2.5.

Along the same lines, the periodic table of the elements is a tabular representation
of a taxonomic classification encoding relations such as

lithium is_a alkali metal,

fluorine is_a halogen,

and so on. But chemists also recognize other relationships that exist between the elements, which can be represented in a corresponding domain ontology. For example,
from the Chemical Entities of Biological Interest (ChEBI) ontology:

fusicoccin has_role toxin,

water has_role amphiprotic solvent,

atom has_part electron,[5]

and so on.

Definition, Taxonomy, Ontology

So far we have stressed the role of taxonomies (*is_a* hierarchies) and of the representation of other relations (such as *part_of*) among universals as essential features of an ontology. As we shall discuss in more detail in chapter 4, explicit and clear definitions of the terms in an ontology are essential also. Indeed, all ontologies, as we understand them here, consist of (1) a central backbone taxonomy, in which all the nodes of the ontology are linked together via *is_a* relations, together with (2) further relations defined between the nodes of the ontology. In addition, each node consists of (3) a term along with, when necessary, (4) synonyms for the term, and crucially (5) a definition of the term that makes use of the Aristotelian genus and differentia structure.

Definitions are perhaps the most important component of ontologies, since it is through definitions that an ontology draws its ability to support consistent use across multiple communities and disciplines, and to support computational reasoning. Definitions also constrain the organization of the ontology. Simply put, every term in an ontology (with the exception of some very general terms) must be provided with a definition, and the definition should be formulated through the specification of how the instances of the universal represented by the relevant term are differentiated from other instances of the universal designated by its parent term.

The resultant regimentation makes possible simple inferences. For example, from

COPI vesicle coat is_a protein complex,

and

protein complex is_a macromolecular complex,

we can infer that

COPI vesicle coat is_a macromolecular complex.

From

MRI image output preceded_by MRI test,

and

MRI test preceded_by referral,

we can infer that

MRI image output preceded_by referral.

The power to make such inferences also allows the ontology to be used for error checking by allowing us to test at any given stage in its development whether its definitions and relations are consistent. If we use the ontology to annotate a body of data—to create what is sometimes called a "knowledge base"—then we can use the result also as a tool to detect certain sorts of errors in our data (as described in chapter 8). And because a domain ontology is relatively independent of the specific data-collection goals of specific groups of researchers, it can also be used to advance the sharing of data among multiple communities of researchers, and to support retrieval of data that might otherwise be masked by the use of labels that are obscure or undefined.

Top-Level Ontology

The use of domain ontologies is becoming increasingly common in many branches of science in reflection of the increasing need to use computers for the handling of scientific data. However, the very success of the ontology-based approach has brought with it the Tower of Babel problems discussed in the Introduction. Multiple groups of researchers are creating incompatible domain ontologies focused on their specific local needs, resulting in new information siloes, with all the problems of inaccessible and nonsharable data, and nonoptimal use of resources, that this brings.

It is above all to counteract this tendency that we promote a strategy focusing on the need to define, first, those terms in an ontology which refer to universals—terms which together form the shared vocabulary of scientists in the relevant domain—and to use these terms as our starting point for developing definitions of terms representing the multiple different sorts of lower-level universals and of defined classes needed for application purposes by different sub-communities of scientists. We believe that insisting that definitions for ontology terms be created in this way can help to ensure the consistent development of ontologies across multiple communities of researchers in much the same way that insisting that all medical students share a common knowledge of the basic biomedical sciences helps to ensure that they are able to communicate with each other as they acquire different sorts of specialist knowledge.[6]

But all Aristotelian definitions need a starting point—some parent (genus) in relation to which the child (species) term can be defined, and this starting point is, given what was said earlier, in every case the root node of the ontology in question.[7] But how can we ensure that the many different ontologies that we will need for the many different domains will be developed in consistent fashion? How can we ensure that the different ontology-building communities will employ comparable and mutually understandable definitions? The answer, we believe, is to ensure that all domain ontologies share a common top-level formal ontology in terms of which their respective root nodes can be defined. This common top level would in effect provide the shadow parent terms needed in order to initiate the process of creating Aristotelian definitions, and thereby provide the common overarching framework needed for consistency.

Where a *domain ontology* is constructed as a representation of a basic set of universals pertinent to a single scientific domain, a *top-level* ontology is a highly general representation of categories and relations common to all such domains. For example, if *cellular division* is a universal in one domain ontology, and *cancer development* is a universal in another domain ontology, then a top-level ontology will include a category such as *process*, which subsumes both of these. Whereas domain ontologies help to integrate and make available information pertinent to a given area, top-level ontologies help to integrate and organize information across different domains.

Our proposal is that domain ontologies be constructed by downward population, using the mechanism of Aristotelian definitions, from a common top level. Experience has shown that in the absence of this common top level, the domain ontologies developed by different groups—for example, the anatomy ontologies developed by different communities of mouse, rat, or human biologists—will be developed in relative isolation from each other. This will make the sharing of annotated information based on these ontologies a difficult and time-consuming task and it will contribute further to the problem of data siloes. Use of a shared top-level by different groups of ontology developers will also increase the flexibility of training; it will allow more effective governance and quality assurance of ontology development; and it will promote the degree to which multiple different groups of ontology developers and users can inspect and critique and help to improve their respective bodies of work.

Semantic Interoperability

One central goal of the annotation of data using ontologies is to enable what is called "semantic interoperability" between heterogeneous computer systems, defined as the ability of two or more such systems to exchange information in such a way that the meaning of the information generated by any one system can be automatically interpreted by each receiving system accurately enough to produce results useful to its end users. The use of a common top-level ontology for describing the data generated by information systems increases the likelihood that these conditions will be capable of being met. It can also assist in advancing the degree to which information systems that use it can support formal reasoning. Thus, where a domain ontology assists in organizing the data of a particular domain to make that data understandable, accessible, and computer analyzable, a top-level ontology assists in organizing the data of multiple domain ontologies in a way that promotes the degree to which the information systems using these ontologies will be semantically interoperable.

Choice of Top-Level Ontology

There are already a number of top-level ontologies in existence that are being used in just the way described here. For example, domain ontologies that comprise the Open Biomedical Ontologies (OBO) Consortium are now using Basic Formal Ontology (BFO) as the standard top-level ontology to assist in the integration of biomedical data and information derived from a variety of different sources. Other examples of top-level ontologies include the Descriptive Ontology for Linguistic and Cognitive Engineering (DOLCE) and the Standard Upper Merged Ontology (SUMO). We will provide a detailed introduction to BFO in chapters 5 and 6. One major advantage of the BFO framework is that over 130 public-domain ontologies, thus far primarily in the biological and biomedical domains, have been developed on the basis of BFO, so that the data annotated using terms derived from the BFO framework is orders of magnitude more comprehensive, both in terms of variety of domains and in terms of size of the datasets involved, than either DOLCE or SUMO.

Both DOLCE and SUMO have their advantages, however. DOLCE is defined as a "Descriptive Ontology for Linguistic and Cognitive Engineering."[8] From the point of view of quantity of users, it is a very successful upper-level ontology, and it has been applied in a number of projects in biology and social science. DOLCE and BFO in fact grew out of a common philosophical orientation, and thus BFO overlaps with parts of DOLCE's top level. In contrast to BFO, which focuses on the universals in reality, DOLCE relies on an ontology of possible worlds, and thus includes in its coverage domain putative objects of mythology and fiction. SUMO, too, has proved to have considerable value as an upper-level ontology for certain purposes.[9] It should be noted, however, that SUMO is not a top-level ontology in the sense that this term is used here. This is because it contains, for example, biological terms ("protein," "crustacean," "body-covering," "fruit-Or-vegetable"), and this means that it cannot cleanly support the strategy of downward population that has proved so useful to scientists in the case of BFO.

Application vs. Reference Ontology

Table 2.1 provides a summary of the different meanings of the term "ontology" that have been surveyed thus far. Where the distinction between domain ontologies and top-level ontologies has to do with the generality of subject matter, we now need to introduce also a distinction between *application ontologies* and *reference ontologies*, which has to do with the goal or purpose for which an ontology is designed.

Table 2.1
Review of three meanings of "ontology"

Philosophical Ontology

• The study of what is, of the kinds and structures of objects, properties, events, processes, and relations in every area of reality (metaphysics). Results in ontologies, descriptions, or theories of what exists, as representational artifacts.

• Has roots in ancient Greece in the work of philosophers such as Parmenides, Heraclitus, Plato, and Aristotle

• Example: the Porphyrian Tree

Material or Domain Ontology

• A structured representation of the entities and relations existing within a particular domain of reality such as medicine, geography, ecology, or law

• A graph-theoretic structure whose nodes are linked by the subtype relation (thereby forming a taxonomy) and by other relations

• Goal: to support knowledge sharing and reuse

• Examples: Gene Ontology (GO), Foundational Model of Anatomy (FMA), Environment Ontology (EnvO), Chemical Entities of Biological Interest (ChEBI), and many others.

Formal or Top-Level Ontology

• Upper-level ontology that assists in making communication between and among domain ontologies possible by providing a common ontological architecture

• Goal: the calibration of interoperable domain ontologies into larger networks

• Examples: Basic Formal Ontology (BFO), Descriptive Ontology for Linguistic and Cognitive Engineering (DOLCE), Standard Upper Merged Ontology (SUMO)

An *application* ontology is an ontology that is created to accomplish some specified local task or application. For example, the Situational Awareness and Preparedness for Public Health Incidents Using Reasoning Engines (SAPPHIRE) information system utilizes an application ontology that classifies unexplained illnesses that exhibit flu-like symptoms and sends this information to the Centers for Disease Control and Prevention.

A *reference* ontology, by contrast, is an ontology that is meant to be a canonical, comprehensive representation of the entities in a given domain that is developed to encapsulate established knowledge of the sort that one would find in a scientific textbook. The Foundational Model of Anatomy (FMA), Gene Ontology (GO), Cell Ontology (CL), and Protein Ontology (PRO) are examples of reference ontologies in this sense.

Portions of multiple existing reference ontologies can be reused in an application ontology, which will typically also contain new ontology content created to address its specific local purpose.

The OBO Foundry (http://obofoundry.org) initiative has embraced the goal of creating a library of semantically interoperable reference ontologies devoted to representing different domains of scientific investigation in such a way that elements of this library could be used and reused for particular application ontology purposes.

Conclusion

In this chapter, we have introduced some basic distinctions between kinds of ontologies. We have explained what a taxonomy is, and discussed the central role that taxonomies play in ontologies. Indeed, it is not going too far to say that an ontology just *is* a very sophisticated type of taxonomy. At this point, our survey of the basic theoretical components needed to make sense out of ontology design is complete. For most ontology designers, the goal will be to build a functional domain ontology, and in the next two chapters we turn to some specific recommendations that are applicable to this process.

Further Reading on Top-Level and Domain Ontology

Simons, Peter M. *Parts: A Study in Ontology*. Oxford: Oxford University Press, 1997.

Smith, Barry. "Ontology." In *Blackwell Guide to the Philosophy of Computing and Information*, ed. Luciano Floridi, 155–166. Oxford: Blackwell, 2003.

Further Reading on Taxonomy and Classification

Jansen, Ludger. "Classifications." In *Applied Ontology: An Introduction*, ed. Katherine Munn and Barry Smith, 159–172. Frankfurt: Ontos Verlag, 2008.

Smith, Barry. "The Logic of Biological Classification and the Foundations of Biomedical Ontology." In *Invited Papers from the 10th International Conference in Logic Methodology and Philosophy of Science*, ed. Dag Westerståhl, 505–520. London: King's College Publications, 2005.

3 Principles of Best Practice I: Domain Ontology Design

In our Introduction, we articulated the problem of managing scientific information in a way that allows combinability and comparability, and we discussed ontology as a proposed general solution to this problem. In chapters 1 and 2, an ontology was defined as a representational artifact whose representations are intended to designate universals, defined classes, and the relations among them. We also introduced some distinctions among different kinds of ontologies, and introduced the idea of a taxonomy as the central component of an ontology. In light of all of this, the problem of designing an ontology is the problem of designing a formalized representational artifact, including a taxonomic hierarchy as backbone, whose representations (terms) designate universals, defined classes, and the relations between them. In this chapter and the next we will discuss what this process looks like in practice, focusing on considerations and principles geared toward the design of domain reference ontologies useful in supporting scientific research. Issues to be considered in this chapter include: subject matter and scope of a domain ontology, as well as the first steps that one should take in designing the ontology itself.

General Principles of Ontology Design

We will first set forth principles specifying the general attitude or outlook to be kept in mind when designing an ontology. Our position is that a good ontology will be one that is designed in such a way as to respect these principles and that, indeed, respecting these principles will be part of what makes an ontology a good one.

1. Realism

We have already discussed our commitment to realism in chapter 1. In general, "realism" can be defined as a philosophical position according to which reality and its constituents exist independently of our (linguistic, conceptual, theoretical, cultural) representations and can be known, for example, through perceptual experience and through application of the scientific method. The goal of science, from this realist (and, we believe, commonsensical) perspective, is to discover truths about reality.

Realism in ontology is based further on the idea that with the aid of science we can come to know the general features of reality in the form of universals and the relations between them. This realist approach has a number of general consequences. First, it implies that ontologies are representations of reality, not of people's concepts or mental representations or uses of language. Certainly an ontology of, for example, cognitive psychology or linguistics might have concepts or mental representations or uses of language within its subject matter. But then the latter would be treated as parts of reality in a way exactly analogous to what is the case in, for example, an ontology of astrophysics or plant development.

Many parts of science pertain to entities, such as chemical elements or prokaryote cells or Paleoproterozoic rocks, which existed long before the first human beings. Other parts of science pertain to entities—for example, in the domains of law or economics—which exist as a result of human thought and activity. Ontological realism applies equally to all branches of science, taking the view that, for example, collateralized debt obligations are no less real than electrons and planets.

2. Perspectivalism

The goal of science is not merely to discover truths about reality. Its goal is to develop theories that are as accurate, as broad ranging, as predictive, as explanatory, as logically coherent, and as well tested as possible. Unfortunately these goals—and a number of other goals that are also found attractive, such as maximal consistency with common sense—seem not to be simultaneously realizable. To cope with this fact, we embrace a doctrine of *perspectivalism*.

Perspectivalism flows from the recognition that reality is too complex and variegated to be embraced in its totality within a single scientific theory. It amounts to the principle that two distinct scientific theories may both be equally accurate representations of one and the same reality.

This does not mean, of course, that all representations created by scientists are of equal value. A view according to which fish are mammals would clearly be of less value than a contrary view, because it would be less accurate to the facts of reality. But there are nonetheless many different representations that are equally good (true, veridical) representations of some given portion of reality precisely because they capture different features of this reality. The most straightforward examples of different but equally legitimate perspectives regarding the same domain of reality have to do with the phenomenon of granularity. Simply put, it is equally legitimate to examine living organisms from a perspective of molecular biology as from a perspective that takes into account anatomy and physiology at the level of organs and organ systems. It is equally legitimate to examine human behavior from the perspective of the physics of the human sensorimotor system as from the perspective of economic incentives. Each of the mentioned perspectives can yield contributions to our knowledge of reality that are accurate to the reality at hand.

The implications of perspectivalism for ontology are that the irreducibility of different perspectives should be respected also in the design of ontologies. Ontology developers should not seek to represent all portions and features of reality in a single ontology, but should seek, rather, a modular approach, in which each module is maintained as far as possible by experts in the corresponding scientific discipline.

3. Fallibilism

Fallibilism involves commitment to the idea that, even though our current scientific theories are the best source we have of statements that are candidates for expressing truths about reality, it may nevertheless be the case that some of these statements are false. Reality exists independently of our ways of understanding it scientifically, and experience tells us that even our best current theories may be subject to correction. Thus while the realist holds that our experiences, ideas, and scientific theories are about reality—that they form in totality a representation, a map or picture, of reality— this does not mean that all elements of this map are correct; some elements may misrefer, some may fail to refer at all.

Our map of reality is at any given stage always only partial: reality is never revealed to scientists in all its totality. And because our representation is continually expanding as we learn and discover more about what exists on the side of reality, what we believe today, based on what we have learned about those facets of reality to which we have so far gained access, is sometimes undermined by what we learn tomorrow about facets of reality hitherto unappreciated.

The process of correction of our map of reality is itself subject to multiple different sorts of setbacks and changes of direction, some (few) of which may be radical in nature (two outstanding examples being the Copernican and Darwinian revolutions in physics and biology). Through all of these changes, however, and even through the most radical of scientific revolutions, major referential elements of this map remain intact. Scientists erred in believing that the sun rotates around the earth; but upon correcting this error they continued to use terms like "sun" and "earth" to refer to the very same entities as before. Something similar applies to general terms like "atom," "star," "organism," "cell," and "planet." While our beliefs about these entities have changed with time, these terms themselves have to a large degree preserved their reference through such changes. At the same time, however, the fallibilist accepts that in regard to general terms, too, our scientific knowledge is subject to being overturned through time through new empirical discoveries, as, for example, in the already mentioned "phlogiston" case.

Some specific implications of fallibilism for ontology design in support of scientific research include the following:

3a. That every ontology must have sophisticated strategies for keeping track of successive versions of the ontology.

A new version of an ontology becomes necessary when errors in current scientific theory about the domain are discovered and corrected and when new information relevant to the domain is discovered. Users of the ontology need to be able to keep track of such changes.

3b. That every ontology needs to have a tracking service for its users that will allow them easily to point out errors and gaps in the ontology and to have their submissions to this service addressed in a timely manner.

Like science itself, ontology design is an ongoing collective enterprise in which errors can be detected and avoided by the input and testing of multiple individuals.

4. Adequatism

There is a widespread tendency in philosophical circles to view the goal of philosophy in reductionist terms. On this view the job of the philosopher is to explain complex phenomena by reducing them to simpler and more fundamental components, here drawing on the astonishing successes of modern physics. *Adequatism* is the opposite tendency, which holds that the entities in any given domain should be taken seriously on their own terms and that room must be made in our set of theories of reality for all of the different sorts of entities that reality contains, at all levels of granularity.

For the adequatist all scientific disciplines are prima facie of equal worth in providing representations of what exists in reality. Just as an ontology of physics is about, for example, atoms and subatomic particles, and an ontology of chemistry is about chemical elements and compounds and the associated reactions, so an ontology of biology will include representations of the universals and defined classes at various levels of granularity from molecules and cells, to organs and organisms, and from there to populations and ecosystems. The goal of ontology as viewed by the proponent of adequatism is to do justice to the vast array of different kinds of entities that exist in the world, rather than ignoring these or those specific kinds of entities or attempting to explain them away.

It is the adequatist view of ontology that is defended in what follows. Suppose, for example, that one needs to create an ontology for a given domain as this domain is described in the textbooks of some given scientific discipline. The ontology should be designed to represent the types of entities described in the textbooks; but it should do this in such a way that it can be linked to other ontologies covering neighboring domains, including domains recognizing entities at different levels of granularity. The implication is that ontologies should not be developed in isolation from each other, but rather always in tandem with ontologies with which they must interoperate.

More generally, an adequate framework for ontology development should allow for entities at multiple levels of granularity (as, for example, in biology, where an adequate general framework must allow for—at least—molecules, cells, organs, organisms, and populations) and for a variety of different kinds of relations between the entities on these different levels.

Additional Principles of Ontology Design

Whereas the foregoing four principles represent general theoretical standpoints about ontology design, the following four are more concrete guidelines concerning the design process itself.

5. The Principle of Reuse

Ontologists should not reinvent the wheel. The first step in ontology development should always be one of examining existing ontology resources in and around the domain of focus in order to identify content that is already available that meets scientific and ontological standards. Ontologies should reuse as far as possible relevant ontological content that has already been created; and even where this content cannot be reused it should be regarded as forming a benchmark that can be used to gauge the adequacy of new content that is created.

Ontologies are designed to support communication between information resources relating to the multiple provinces of reality and to the multiple disciplines which seek to describe them. They can be compared, in this respect, to highway systems. It will very rarely be the case that the correct solution to an ontological problem involves the equivalent of ignoring all the roads that already exist, and creating an entirely new highway system from scratch.

At the same time, however, it must be emphasized that—precisely because ontologists have so often ignored design principles of the sort presented here, and because they have themselves often created new ontologies from scratch—much available ontology content is poor in quality, and so due diligence is required not merely in identifying potential ontologies for reuse, but also in evaluating the ontologies identified (and in some cases recommending that they be excluded from further use).

6. The Ontology Design Process Should Balance Utility and Realism

It is an implication of realism that some representational schemes are better than others because they are better representations of reality. Given that some of the roots of ontology building lie in the field of what is sometimes called knowledge engineering, where highly practical motives predominate, it is often argued that ontologies should be measured not by this global standard of adequacy to reality—a standard adapted from the domain of science as a whole—but rather by their utility to some specific local purpose. From our point of view, however, this focus on local utility is wrongly conceived if it is seen as involving a sacrifice on the side of adequacy to the reality which the ontology is being constructed to represent. For it is this reality—as described in the best current science—which provides the common benchmark that can ensure that ontologies are developed in a consistent fashion. Ontologies can indeed be developed in the absence of such a benchmark, but then when they are used to annotate data, the results will not be combinable—except perhaps through considerable manual effort—with data collected by others in neighboring domains. One lesson from over fifteen years of experience with the Gene Ontology is that the purpose for which an ontology originally is constructed may differ in significant ways from what turn out to be important secondary uses that could not have been anticipated when the ontology was first conceived.

7. The Ontology Design Process Is Open-Ended

The principles discussed so far provide a framework for understanding a crucial further point about ontology design: designing a domain ontology, at least in the scientific domains of primary concern to us here, is but the first step in an open-ended process of continuously maintaining, evaluating, updating, and correcting the ontology, and of adjusting the ontology to neighboring ontologies in order to take account of advances both in scientific knowledge and in our knowledge of ontology and its associated logical and computational technologies.

Realism implies that the central goal of a good ontology in support of scientific research is to represent reality adequately. But it implies also that for scientific domains we are at any given stage almost always in possession of only partial information about the reality at issue. Our strategy thus forces upon us the precept that ontologies should be designed in such a way as to be expandable and amendable through time, and the best practice principles that we will be discussing in what follows are designed to serve this end. Note that this precept is consistent with the fact that there will be practical constraints on the ontology builder resulting from the fact that resources for populating an ontology are limited by economic and other circumstances. For while those branches of the ontology that are associated with the most urgent needs will be developed in greatest detail, the population of such branches will be of greater utility if it is managed within a general framework that can allow coherent population of neighboring branches in the future.

8. The Principle of Low-Hanging Fruit

A final general principle to keep in mind is the following: when designing a domain ontology, begin by identifying those features of the subject matter that are the easiest and clearest to understand and define.

The ontologist should begin, in other words, by gathering the low-hanging fruit on the ontology tree, including what to a human being may seem like trivial assertions (for example, *cell membrane is_a membrane*) but which to the computers who will process the ontology are indispensable. In constructing a domain ontology we start by categorizing the simple universals and relations first. As a general rule, the ontology developer should start by identifying the general terms most commonly used in the beginning chapters of relevant introductory textbooks, and move on from there, step by step, to the representation of more complex entities within the domain. The principles of ontology design that have been surveyed up to this point are summarized in box 3.1.

Overview of the Domain Ontology Design Process

Ontology is a top-down approach to the problem of electronically managing scientific information. This means that the ontologist begins with theoretical considerations of a very general nature on the basis of the assumption that keeping track of more specific information (for example, about specific organs, genes, or diseases) requires getting the very general scientific framework underlying this information right, and doing so in a systematic and coherent fashion. It is only when this has been done that the detailed terminological content of a specific science such as cell biology or immunology can be encoded in such a way as to ensure widespread accessibility and usability. The method to be followed in constructing a domain ontology on the basis of this general starting point can be summarized in the steps outlined in table 3.1.

Box 3.1
General Principles of Ontology Design

1. *Realism*: the goal of an ontology is to describe reality.
2. *Perspectivalism*: there are multiple accurate descriptions of reality.
3. *Fallibilism*: ontologies, like scientific theories, are revisable in light of new discoveries.
4. *Adequatism*: the entities in a given domain should be taken seriously on their own terms, not viewed as reducible to other kinds of entities.
5. *The Principle of Reuse*: existing ontologies should be treated as benchmarks and reused whenever possible in building ontologies for new domains.
6. *The Ontology Design Process Should Balance Utility and Realism*: sacrificing realism to address considerations of short-term utility when building an ontology may be at the detriment of the ontology's longer-term utility.
7. *The Ontology Design Process Is Open-Ended*: scientific ontologies will always be subject to the need for update in light of advances in knowledge; ontology design, maintenance, and updating is an ongoing process.
8. *The Principle of Low-Hanging Fruit*: in ontology design, begin with the features of the relevant domain that are easiest to understand and define, then work outward to more complex and controversial features.

Table 3.1
An outline of the steps to be followed in designing a domain ontology

1. Demarcate the subject matter of the ontology.
2. Gather information: identify the general terms used in existing ontologies and in standard textbooks; analyze to remove redundancies.
3. Order these terms in a hierarchy of the more and less general ones.
4. Regiment the result in order to ensure:
   a. logical, philosophical, and scientific coherence,
   b. coherence and compatibility with neighboring ontologies, and
   c. human understandability, especially through the formulation of human-readable definitions.
5. Formalize the regimented representational artifact in a computer usable language in such a way that the result can be implemented in some computable framework.

Step 1 consists of determining and demarcating the subject matter of the ontology one needs to build. This will involve establishing the nature and scope of the data (for example, experimental or clinical) one needs to annotate, and identifying existing ontology content in relevant domains. The initial survey of the content of the pertinent science should yield provisional answers to the following questions:

• What are the domain universals and relations that need to be represented?

• What are the appropriate domain-specific terms that should be used in representing these universals and relations?

• What levels of granularity of entities are salient for the ontology?

Step 2 is the task of assembling a selection (fifty or so) of the most common highly general terms, some of them taken from relevant ontologies, some from standard textbooks.

Step 3 is the provisional ordering of these terms in a hierarchy of the more and less general and serves as the precursor to step 4. Step 4 consists in working on this hierarchy to ensure coherence, for example, by adding further terms to ensure a complete taxonomical hierarchy for the ontology; and identifying the terms referring to the highest-level universals in the domain in question, which will serve as the root node or nodes of the ontology being developed. It will involve also creating a set of human-understandable definitions for the selected terms, which will include gathering further information concerning the most important domain-specific universals which are subsumed by these highest-level universals, and identifying any relevant terms in neighboring ontologies which will be needed in the formulation of the definitions. Starting with the root nodes and working downward, we attempt to identify successive genera and differentiating characteristics that will need to be included in the definitions of the entities to be included in the ontology; and we adjust our preliminary classification scheme in light of any changes which our definitions dictate.

The process of regimentation is iterative, and will involve successive cycles of reviewing versions of the hierarchy of terms and definitions for logical, philosophical, and scientific adequacy, including consistency and human intelligibility, and also ensuring that the result leaves out no essential elements of the domain.

Once a thorough understanding of the domain has been established in this way, step 5 is the task of iterative encoding of the ontology through logical formalization. This is achieved by transforming the natural language definitions for the terms contained in the ontology into a format that is computer usable using an ontology editing tool.

While the process of five steps is top down in nature, working from the highly general to successively less general terms in the ontology, in practice it will involve a great deal of cycling via feedback loops between the successive steps. In the following sections, we will discuss the demarcation and information-gathering processes in more detail. In chapter 4 we will address the issue of regimentation, and we will return to issues of encoding in chapter 8.

Explicitly Determine the Subject Matter of the Domain Ontology

The first step in constructing a domain ontology is to determine explicitly the intended scope of the ontology, which is to answer the question: "what part of reality is this ontology an ontology of?" Providing an explicit statement of this scope will serve to indicate both what is to be included in and what is to be excluded from the intended ontology. For example, the documentation for the Foundational Model of Anatomy (FMA), an ontology of human anatomy, describes the ontology as "strictly constrained to 'pure' anatomy, i.e., the structural organization of the body."[1] This statement makes it clear which terms are candidates for inclusion in the FMA, but also which terms to exclude: those relating, for example, to functional anatomy, or surgical anatomy. The specification of scope will also indicate the level or levels of granularity of reality against which the ontology is calibrated. Will this be *multicellular organisms*, or *organs*, or *cells*, or *cell components*, or *molecules*? Or will it perhaps be *entire populations of organisms*? Or will it be some combination of levels, as in an ontology that deals with cell signaling and thus needs to represent, for example, both cells and signaling pathways?

Domain and Top-Level Ontologies

We have seen that for purposes of successfully managing scientific information in the long term, the root node or nodes of a domain ontology should be defined in terms of some highly general universal taken from a domain neutral ontology such as BFO. This will help to ensure that the ontology is built using a high-level ontology architecture that is shared with other ontologies.

If, for example, the relation *part_of* is asserted in a given formal ontology as being transitive (so that if *x part_of y* and *y part_of z*, then *x part_of z* will hold), then in a domain ontology built on its basis, for example, in the domain of anatomy, we will be able to use this feature of the parthood relation to infer from *finger part_of arm* and *arm part_of body*, to *finger part_of body*.

Similarly, if a top-level ontology contains distinct representations for continuants (three-dimensional entities that continue to exist through time, such as planets and molecules) and occurrents (entities that occur, which means that they are spread out not just in space but also in time, such as a baseball game or the movement of a planet in its orbit), then all domain ontologies defined on its basis will be required to respect this same distinction among the entities it represents.

In these and a series of related respects the top-level ontology helps to ensure the correctness of ontology construction at lower levels. If an ontology uses *part_of* but contains assertions that contradict transitivity, then these assertions can be flagged as being in need of manual checking. If an ontology recognizes the distinction between things and processes, then problem cases—for example, terms such as "gene mutation," which are ambiguous as between thing and process readings—can be identified in advance and warnings issued requiring developers to subject such terms to additional manual inspection. BFO was designed to play this kind of role in the process of domain ontology design and quality assurance.[2]

For these reasons it is important, at the outset of designing a domain-specific ontology, to consider what top-level ontological categories and relations might apply to the domain at hand, and to select a top-level ontology representing sufficient and sufficiently clear categories and relations to handle the basic kinds of entities to be found in the domain in question. It is important to note that, by definition, a top-level ontology should be domain neutral. Thus it should not contain representations of relations and universals that are specific to any given domain. It will thus, in comparison with many of the domain ontologies defined in its terms, be very small. The ontological content pertinent to each specific domain is then added to that of the top-level ontology through the process of downward population.

Relevance

The task of determining what portion of reality a domain ontology should represent involves also addressing the problem of determining what and how much of the existing data and information about a given domain should be included in the ontology. It can be summarized as the problem of determining what is *relevant* for the ontology, a matter to be determined (1) by the current state of science, and thus by the structure of the corresponding portion of reality, (2) by the degree to which existing ontologies in neighboring areas can be relied upon in supporting the given ontology's development, and (3) by the practical goals the ontology needs to satisfy.

What is objectively relevant to the Cell Ontology (CL), for example, is determined by the nature of cells themselves, what they are, what processes they characteristically initiate or are involved in, and so forth. The CL taxonomy of immune cells is created on the basis of information about the protein molecules expressed on cell surfaces; representations of relevant types of molecules are drawn from the Protein Ontology (PRO), to create definitions such as the following:

lymphocyte of B lineage = def. lymphocyte and (has_plasma_membrane_part some CD19 molecule) and (lacks_plasma_membrane_part some CD3 epsilon)

Or in other words: a lymphocyte of B lineage is a lymphocyte that has CD19 molecules on its plasma membrane but does not have CD3 molecules on its plasma membrane. Here "lymphocyte" is a higher-level term defined in CL, "CD19 molecule" and "CD3 molecule" are defined in PRO, and "plasma membrane" is defined in the Cellular Component branch of the Gene Ontology.

Connections between cells and proteins are handled by building links between relevant ontologies in such a way that the information compiled in each of these ontologies is brought together in ways useful for reasoning and integration. By this means we avoid also some of the dangers of silo formation—for example, where those interested in cells feel tempted to develop their own local ontology of protein surface markers, an ontology that would fail to interoperate with other protein information resources. Ensuring that the corresponding domain ontologies have been structured on the basis of the same top-level ontology from the start makes it easier to bring them into alignment in the needed way.

The task of determining what needs to be represented in an ontology will also depend on the practical goals the ontology needs to satisfy. All ontology development (like all of science) is to some degree opportunistic: which parts of an ontology are developed first, or in the greatest detail, will often depend on available funding, and such funding will often be bound to purpose. Goal-oriented human activities bring some entities into focus and leave others in the background. If our job is to support the scientific investigation of a hypothesis relating to, say, fetal diseases involving lymphocytes of B lineage, then we will first identify existing ontologies with relevant content. But our investigation may require the development of an entirely new ontology focusing strictly on specific areas—for instance on interactions between these and those specific kinds of cells and molecules in these and those kinds of patients undergoing these and those kinds of treatment.

These ways in which purpose can determine ontology content reflect the distinction introduced in chapter 2 between *reference ontologies* and *application ontologies*. A reference ontology is a representational artifact analogous to a scientific theory, in which maximizing expressive completeness and adequacy to the facts of reality are of primary importance. An application ontology is a representational artifact that is designed to assist in the achievement of some specific goal. Reference ontologies will be constructed and structured primarily on the basis of the established content of a scientific discipline. Application ontologies will be constructed and structured primarily in terms of what is relevant to some specific goal. To be successful in the long term, however, application ontologies should to the maximal possible extent make use of portions of reference ontologies as their starting points. Development of application ontologies thereby can also benefit work on reference ontologies, for example, where terms created within the framework of the former are discovered to be of general scientific relevance, for example, then these terms will be promoted to a level where they form part of a reference ontology available for more general use.

Granularity

One subpart of the problem of determining relevance is the issue of determining the appropriate granularity of the entities to be represented in an ontology. The issue of granularity arises because things in reality, and also the parts of things, come in many different sizes and possess varying degrees of complexity. There is a continuum stretching from subatomic particles, atoms, and molecules, through ordinary objects such as animals, rocks, and tables, on to ecosystems, planets, solar systems, galaxies, and ultimately the universe itself. There is an analogous continuum also in the realm of processes unfolding in time, stretching from milliseconds to years to geological epochs. Things and processes can be identified at all of these different granularity levels, and as we move up to successively coarser grains we encounter entities that exhibit features not found at lower levels—a phenomenon referred to by philosophers under the heading of "emergence." The problem of granularity for ontology design is the problem of deciding the prototypical sizes and complexities of entities that are to be represented in a given domain ontology. Should an ontology of mountains include representations of the types of molecules of which the mountains are composed? Should an ontology of stages in the plant life cycle include stages of growth of individual leaves? When setting out to design an ontology, the choice of root nodes will determine, in part, the level or levels of granularity that will form part of the ontology's coverage, but this determination will be influenced primarily by the needs of users of the ontology—for example, in reflection of the degree to which finer gradations of taxonomy enable the recording of differences in data of a practically useful sort.

The Problem of Nonexistents

Once the domain or scope of an ontology has been decided on, what is needed is a systematic survey of the content of the established science relevant to this domain. This means primarily surveying the current content of authoritative textbooks and of salient terminologies. Thus ontologies relate primarily to the use of general terms in established sciences. Ontologies may in areas such as chemistry be used to represent types of entities which do not exist—for example, molecules not yet synthesized—but in general the rule is that ontologies should consist of representations only of those types for which we have good evidence that instances exist (and, by extension, only of those defined classes for which we have good evidence that there are members).

Very occasionally, ontologies may need to be developed to support research in areas still subject to dispute among different groups of scientists and thus not belonging to established science. (Recall, again, the case of the "Higgs boson.") We prefer to see such ontologies as provisional in nature, to be promoted to the ranks of ontologies proper only when the disputes in question have been settled. The methods for the creation of such provisional ontologies will then be essentially the same as those outlined here, but will apply the process of term selection not to established textbooks but, for example, to journal articles produced by some subset of the disputing partners. The results of such provisional ontology development will then also be provisional. They will be able to be added to existing ontology content and treated like other ontologies only when the disputes in question have been resolved.

Conclusion

In this chapter we have introduced some general principles of ontology design and provided an overview of the two initial phases of the ontology building process—namely, *demarcating the domain of the ontology* and *gathering information about the domain*. In the following chapter, we will discuss the third step of the ontology building process: regimentation, which deals in greater detail with issues of terminology selection, definition, and classification.

Further Reading on Relevance, Perspectivalism, Granularity, and Adequatism

Bittner, Thomas, and Barry Smith. "A Theory of Granular Partitions." In *Applied Ontology: An Introduction*, ed. Katherine Munn and Barry Smith, 125–158. Frankfurt: Ontos Verlag, 2008.

Hill, David P., Barry Smith, Monica S. McAndrews-Hill, and Judith A. Blake. "Gene Ontology Annotations: What They Mean and Where They Come From." *BMC Bioinformatics* 9 (suppl. 5) (2008): S2.

Kumar, Anand, Barry Smith, and Daniel Novotny. "Biomedical Informatics and Granularity." *Functional and Comparative Genomics* 5 (2004): 501–508.

Masci, Anna M., Cecilia N. Arighi, Alexander D. Diehl, Anne E. Lieberman, Chris Mungall, Richard H. Scheuermann, Barry Smith, and Lindsay G. Cowell. "An Improved Ontological

Representation of Dendritic Cells as a Paradigm for all Cell Types." *BMC Bioinformatics* 10 (February 2009): 70.

Rector, Alan, Jeremy Rogers, and Thomas Bittner. "Granularity, Scale and Collectivity: When Size Does and Does Not Matter." *Journal of Biomedical Informatics* 39 (2006): 333–349.

Smith, Barry. "Ontology (Science)." In *Ontology in Information Systems: Proceedings of the Fifth International Conference* (FOIS 2008), ed. C. Eschenbach and M. Gruninger, 21–35. Amsterdam: IOS Press, 2008.

Smith, Barry, and Werner Ceusters. "Ontological Realism: A Methodology for Coordinated Evolution of Scientific Ontologies." *Applied Ontology* 5 (3–4) (2010): 139–188.

Smith, Barry, et al. "The OBO Foundry: Coordinated Evolution of Ontologies to Support Biomedical Data Integration." *Nature Biotechnology* 25 (11) (November 2007): 1251–1255.

4 Principles of Best Practice II: Terms, Definitions, and Classification

We assume that, following the recommendations advanced in chapter 3, the appropriate scope of the ontology has been determined and the relevant domain information assembled. We assume also that the ontology builder has created a draft list of terms and associated these with a first draft set of definitions and a provisional *is_a* hierarchy. The next step is to use this list of terms to regiment the domain information in a systematic way, while at the same time allowing an improved understanding of the domain to generate improvements in the list of terms. The goal is to develop a representational artifact that is as logically coherent, unambiguous, and true to the facts of reality as possible.

There are three major facets of regimentation for domain ontologies: terminological, definitional, and location in an *is_a* hierarchy. We will treat each of these issues in turn, though the reader should bear in mind that there is a large degree of overlap and interdependency between the three sets of issues.

Principles for Terminology

Gather and Select Terminology

In chapter 3 we suggested that a good starting point for ontology building is to create a set of terms selected from the most commonly used terms in standard textbooks and in relevant domain ontologies. A first and indispensable step in any ontology development project is to perform due diligence in identifying existing ontology content that is relevant to the task in hand, and to evaluate this content for potential reuse, drawing on tools for searching ontologies such as the NCBO Bioportal (http://bioportal.bioontology.org).

The resultant list of words (or better: of common nouns and noun phrases) forms the first draft of what we can think of as a terminology for the domain in question. Such a terminology may have utility already for human beings, for example, in supporting consistent use of language in exchanging information. For us, however, it has a more ambitious purpose, which is to enable the scientific information with which it is associated to be incorporated into the specific type of computer-based representational artifact that is an ontology, and for this a special sort of terminology will be needed.

The Gene Ontology (GO), by far the most successful ontology to date, was described by its creators as a "controlled vocabulary" to be used for regimenting the ways in which information about gene products in different model organisms is described. The problem it was designed to address is common across the whole of science: where multiple disciplinary groups are involved in the study of some scientific phenomenon of interest each will likely have its own idiosyncratic vocabulary. The problem is that there are *too many terms* for purposes of successful information exchange across disciplines. The GO provided a strategy to solve this problem by disseminating a set of "preferred terms" for use in describing attributes of gene products in a species-neutral way. Preferred terms are then used systematically by literature curators to describe experimental data appearing in published papers. These data then become more easily retrievable and combinable, in ways that overcome the problems caused by multiple conflicting vocabularies.

The success of the GO is due in large part to the fact that the influence of its creators was such that they were able to establish their chosen preferred labels as attractors for a large core of users in each of a variety of multiple interacting disciplines studying a variety of different species of organisms. To replicate this success, ontology builders today need to find a way of selecting terms that are as close as possible to the actual usage of a large fraction of those working in the relevant field without alienating those working in this field whose established terminologies involve the use of different terms. This goal can be achieved, in part, through disseminating the chosen preferred labels by using them in the curation of large bodies of data useful to the wider community, and—again following a practice pioneered by the GO—incorporating community-specific "synonyms" into the ontology alongside the preferred labels. Three principles of terminology construction can then be gleaned initially from the experience of the GO:

1. Include in the terminology terms used by influential groups of scientists for the most important types of entities in the domain to be represented.

2. Strive to ensure maximal consensus with the terminological usage of scientists in the relevant discipline. This may well involve working with domain experts, for instance in negotiating terminological compromises.

3. Identify areas of disciplinary overlap where terminological usage is not consistent. Look for and keep track of synonyms for terms already in the terminology list from these areas.

This strategy alone will work in cases where overlapping disciplines differ merely in their choice of words for the representation of identical entities. Where the terminologies employed by distinct disciplines in such overlapping areas differ in more substantial ways, more complex strategies need to be employed. Two ontologies may, for example, deal with the same phenomena, but at different levels of granularity (for example, molecule and cell); or they may differ in that one ontology deals with objects while another deals with processes; or one may deal with objects, while the other deals with images of objects.

In such cases multiple ontologies must be developed (or multiple branches of a single ontology), and the corresponding terms connected together through relations and through corresponding definitions and axioms. These are workable strategies because we are dealing with areas of *established* science, where we can assume that the disciplines in question will be consistent with each other as concerns their scientific content. Often it will be possible to formulate mapping rules—analogous to, for example, rules for conversion between different systems of scientific units—which allow assertions formulated using the terms from one discipline selected as synonyms in an ontology to be converted into assertions formulated using the terms selected as preferred labels.

What is at all costs to be avoided is the creation of entirely new expressions as preferred labels in ontologies to represent entities with which domain experts are already familiar under established names. Similarly, the ontologist should avoid using familiar terms with new and different meanings. To avoid confusion both in the encoding of information into the ontology and in the interpretation of such information by end users, the terminological choices of domain ontology builders should be as respectful as possible of the current terminology, usage, and practice of contemporary domain experts and of potential users. This leads to a fourth principle for terminology construction, which echoes the principle of reuse from chapter 3.

4. Don't reinvent the wheel. In term selection, stay as close as possible to the usage of actual domain experts. In terminology construction and ontology design, make use of as many existing resources (terminologies and ontologies) as possible.

Formatting Terminology

5. Use singular nouns.

The terms in an ontology should as far as possible have the grammatical form of singular nouns or singular noun phrases.

Two sorts of reasons support adoption of this convention. First (and this will be a common refrain in what follows when we deal with recommendations about syntax and terminology), it is crucial that some syntactical standard, some rule of the road, be adopted and adhered to by all of those involved in ontology building in order to synchronize the multiple such efforts running in parallel at any given time. To see what happens when this rule is not followed, consider, for example, the case of MeSH,[1] whose hierarchy implies *is_a* relations such as

communism is_a political systems,

political systems is_a social sciences,

social sciences is_a behavioral disciplines and activities,

behavioral disciplines and activities is_a topical descriptor,

and so forth. Mixed use of singular and plural nouns may be perfectly appropriate for purposes such as the construction of library catalogs; it causes problems, however, when compiling information in a form that will be reasoned over.

The singular noun rule has been well tested in practice, and yields a simple and cost-free form of synchrony. There is also a principled reason for insisting that all terms in an ontology should take the form of singular nouns. This is because each such term is intended to refer not to some plural or collective entity, but rather either to a universal or to a defined class. In either case, its reference will be singular. There is only one universal *organism*, even if it has many instances, and there is only one defined class *cause of traffic accident*, even if it has many and diverse members.[2]

6. Use lowercase italic format for common nouns.

Along the same lines as principle 5, we recommend that when preparing ontology content for review by human beings *lowercase italic letters* be used for terms referring to universals or classes (this recommendation being based in part on the fact that initial capital letters are normally used in English to indicate proper names, which are names of instances ("Tom," "Seattle," "Jupiter"). Thus *cat*, not "Cat" or "CAT," and *eukaryotic cell*, not "Eukaryotic Cell" or "EUKARYOTIC CELL."

Some ontology editing programs require the use of underscore (eukaryotic_cell) or single quotation marks ('eukaryotic cell') or camel case (eukaryoticCell) in order to allow the beginnings and endings of noun phrases to be identifiable by the computer. Whichever traffic rule is chosen in this respect, the main goal is to ensure that the convention is consistently adhered to—again for reasons of cross-ontology coordination.

7. Avoid acronyms.

Avoid as far as possible the use of acronyms and abbreviations in formulating ontology terms. The rationale for this is that acronyms and abbreviations are too easy to create locally—often, for example, by designers of databases for no reason other than to enable all column headings to fit on a single screen. The half-life of acronyms can be very short, and it is not unusual for those who work with databases (even, sometimes, a database's own creator) to forget what their acronyms originally meant. The goal of ontology, in contrast, is to create standard terminologies that can be employed and relied upon by anyone—in the present and in the future—working in a given discipline. Some acronyms and acronym-involving expressions have in some scientific idiolects become part of the language, as, for example, in terms such as "DNA" or "AIDS," or "ATPase"; they have become in this way safe from the possibility of being reused by new groups of researchers with different meanings. Apart from such cases, however, when selecting a primary label for an entry in an ontology a complete noun or noun phrase should in every case be used.

8. Associate each term in the ontology with a unique alphanumeric identifier.

The identifier is associated with the term in a given version of the ontology. Whenever the ontology is revised and published in a new version, then provided the term in question is not changed in this revision, its identifier can be preserved without change. Identifiers are needed for computer purposes—they will, for example, form the basis of the universal resource identifiers with which ontology terms will be identified in web-based systems. Figure 4.1 is a screen shot of a fragment of the Protein Ontology (PRO) that illustrates the approach we recommend.[3]

At the top of the hierarchy in figure 4.1 is the entry for "amino acid chain." Clicking on the entry will take the user to a human-readable definition of the term, along with other information about it. To the left of the term is its alphanumeric identifier PR:000018263, which uniquely identifies the location of this term in the PRO structure for purposes of computer programming and is used also in the creation of cross-links from other ontologies and databases back to the PRO. The identifier will be associated not merely with the term but also with its unique human-readable definition (for purposes of construction, maintenance, and use of the ontology by human beings), and also with the logically formalized version of this definition.

9. Ensure univocity of terms.

Terms should have the same meaning on every occasion of use. In an ontology, "cell" should refer always to the universal *cell*, "cancer" always to the universal *cancer*, and so on. The principle of univocity in ontology terminology development is difficult to maintain because it is so regularly violated both in ordinary and in scientific (and clinical) language. This occurs, first of all, because of ambiguous expressions, including "cell" itself, which has not only a biological meaning but also (related) meanings in

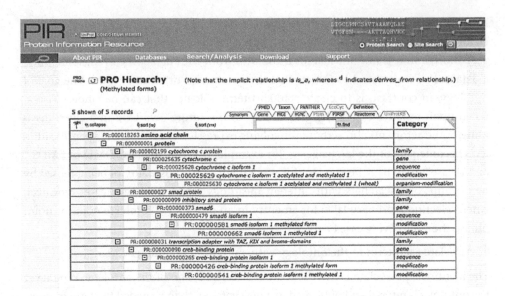

Figure 4.1
Screenshot from the Protein Ontology (PRO) browser containing terms identified by alphanumeric identifiers

relation to, for example, prison cells or cells in a spreadsheet. A more important reason, however, turns on the fact that departures from univocity occur because of the human tendency to use ellipsis in local circumstances (for example, to use "third left hip" to refer to the hip fracture patient in the third bed on the left-hand side of the ward). The reason for insisting upon univocity in the context of ontology design is quite straightforward. If the same term is used in different ways in different contexts, then the humans involved in ontology building are more likely to make errors. Ontologies are of course devised for use primarily by computers, and there the problems of ambiguity are alleviated by the use of unique alphanumeric identifiers for each ontology term. Working hard to avoid departures from univocity is still important, however, since experience shows that such departures are a source of human errors during ontology authoring and maintenance.

It should be noted here that what the principle of univocity specifically says is that every term in an ontology should have exactly one meaning. We do not rule out the presence in an ontology of multiple terms having the same meaning—but this should occur always through the device of declaring one such expression the preferred term, with which synonyms may then be associated according to the terminological needs of the different communities using the ontology.

An example of violation of the principle of univocity is the treatment of the term "disease progression" in the National Cancer Institute [NCI] Thesaurus (version dated August 2, 2004), which offered three different possible interpretations:

(I) Cancer that continues to grow or spread;

(II) Increase in the size of a tumor or spread of cancer in the body;

(III) The worsening of a disease over time. This concept is most often used for chronic and incurable diseases where the stage of the disease is an important determinant of therapy and prognosis.[4]

In definitions (I) and (II) "disease progression" is something that involves only cancer; in definition (III), however, "disease progression" involves the worsening of any disease over time. In the third definition, too, a "disease progression" is identified as a "concept," not as a process. This definition also contains a clause describing how the term is often used. Such information can be included in a comment that is associated with the term in question; for logical reasons, however, it should not be included in the definition itself.[5]

Note that the identified problems still persist in the current (June 30, 2014) version of the NCI Thesaurus, where we have, for example, two terms "cell," defined as meaning "any small compartment" and as "the individual unit that makes up all of the tissues of the body." The former is asserted to be a subtype of "conceptual entity"; the latter of "microanatomic structure."[6]

10. Ensure univocity of relational expressions.

Univocity applies also to the relational expressions used in ontology hierarchies, for example, *is_a* and *part_of*. The early years of ontology development were marked by a phenomenon of "*is_a* overloading" whereby "*is_a*" could mean in different contexts either subclass of, or instance of, or some confused mixture of both.[7] Similarly, "A *part_of* B" was sometimes used to mean that all As are part of some B, all Bs have some A as part, some As have some Bs as part, or again some confusing mixture of all of these.[8] For further details of how these issues are to be resolved, see chapter 7.

11. Avoid mass nouns.

Related to the issue of univocity is a basic distinction between *count nouns* and *mass nouns*. Count nouns, such as "cat," "petal," and "cell" refer to universals whose instances can be counted. Thus it is possible to ask *how many* questions (how many cats are there in this building?, how many petals on this flower?, and so on). Terms such as "water," "tissue," "meat," and "chemical substance" are often used as mass nouns. This means that they are used to pick out or refer to a more or less indefinite quantity of stuff. It is possible to ask *how much* water, meat, or chemical substance there is, for example, in a given container; but not, without further qualification, *how many waters, tissues, meats*. Rather, we ask: "how many *glasses* of water are there?," "how many *pieces* of meat are there?," "how many *liters* of milk are there?," and so on. Now, however, we have replaced the original mass noun with a count noun (more precisely with a count noun phrase) as a means of ensuring that there will indeed be discrete portions of stuff that can be counted.

Certainly there are meaningful sentences involving mass nouns that have not been transformed into count nouns in this way, for instance when a nurse is instructed to store tissue in the freezer or to draw blood from a patient. Reflection reveals, however, that the corresponding transformation is here still being made—even if not explicitly. This is because the relevant amounts or containers are understood. In different contexts, moreover, terms like "blood" may be used to refer not merely to some specific amount of a patient's blood, but to an arbitrary portion or to the maximal portion of blood in a patient's body, and so forth—and "arbitrary portion of blood" and "maximal portion of blood," too, are perfectly acceptable from the point of view of the "avoid mass nouns" principle. A further reason for advancing this principle turns on the ambiguities that arise from the fact that terms like "blood" or "tissue" or "water" or "meat" or "aspirin" are often used to refer to *types*, rather than to particular *portions*, of the substances in question. These ambiguities are of particular importance for ontology builders, since it is precisely the division between types (universals) and instances (particulars), on which ontology is based.

Clearly masses of substances of different types do indeed exist in reality—but on the level of instances they always exist in large or small portions. Thus there is no sugar without there being some determinate portion of sugar; no luggage without there being some determinate number of suitcases and other luggage items. In addition, masses of substances exist on different levels of granularity: thus a mass of body tissue is at one and the same time a collection (mass) of cells.

To summarize: a mass noun such as "tissue" might be used to refer to one or more of the following:

• a portion of stuff within a larger portion of stuff (the tissue within an organ from which a doctor intends to take a sample);

• a discrete (detached) portion of stuff (such as tissue that has been grown independently in order to be placed inside an organ);

• a type of tissue under consideration (lung tissue vs. muscle tissue, healthy tissue vs. cancerous tissue); and

• a maximal or complete quantity of stuff (such as *all* of the tissue comprising the liver).

These different senses of the term "tissue" are involved in quite different theoretical and practical contexts and so it is important to keep them separate for purposes of ontology design. And even if only one such use of a mass noun like "tissue" were selected as the preferred label in an ontology, the mentioned ambiguities would still lead to problems of misuse of this term by human beings. It is for this reason that we recommend that mass nouns be avoided entirely when constructing ontologies. Instead phrases beginning with an appropriate prefix (such as "portion of," "maximal portion of," and so on) should be adopted. This solution has been embraced, for example, by the FMA ontology, which is the leading resource for terms relating (*inter alia*) to different tissue and other body substance types.[9]

To achieve this regimentation, we recommend transforming mass nouns such as "chemical substance" into count nouns by attaching "portion of" or some contextually appropriate equivalent operator to the beginning; thus "portion of chemical substance," "portion of tissue," and so on. Adopting this strategy makes it possible to treat seeming mass nouns as instances of either *fiat parts* or *object aggregates* (see chapter 5). The basic idea though, is that because mass nouns refer to different kinds of entities on different occasions of use, they should be avoided in favor of more ontologically transparent terminology.

12. Distinguish the general from the particular.

Up to this point we have stressed that an ontology is a representation of universals and defined classes. Particular entities—the instances of universals and the members of defined classes—are dealt with, for example, in databases or clinical notes or experimental logbooks. For us, this is a matter of the definition of the word "ontology." Certainly there are some who build ontologies that include an admixture of terms representing individuals—for example, the Standardized Nomenclature for Medicine (SNOMED) includes the term "National Spiritualist Church," which it treats as a subclass of *spiritual or religious belief*.[10] Our reasons for insisting that ontologies should be restricted exclusively to representations of what is general are manifold—but it will be sufficient, for the moment, to mention just one, which is illustrated all too well by the just-mentioned example from SNOMED. Namely, that departure from this principle is often associated with the making of errors: a *church*, however understood (whether as an organization or as a building) is not a special kind of *belief* as SNOMED would have it.[11]

Where an ontology needs to be supplemented by terms representing individuals, then this should be in some separate information artifact—corresponding to the distinction in the Description Logic community between a T-box (for "terminology") and an A-box (for "assertions").[12] The two artifacts can be combined for practical purposes

wherever necessary, forming what some call a "knowledge base." But the result is—
again for definitional reasons—not an ontology, any more than the description or illustration of how a scientific theory has been applied in a specific series of experiments is
itself a scientific theory.

Terms referring to universals and terms referring to instances should be clearly distinguished. For example, the common noun "teapot" as it occurs in a sentence such as
"the teapot is a device for pouring tea" can plausibly be understood as referring to a
type or universal *teapot*. The term "teapot" as it occurs in the sentence "John's teapot
has been stolen" has to be understood as referring to a single particular teapot.[13]

Principles for Definitions

13. Provide all nonroot terms with definitions.

We have addressed the syntactic issues involved in regimenting the terminology of an
ontology, for example by addressing conventions for use of nouns and noun phrases.
An ontology is however above all a semantic artifact—it has to do with regimenting
terms in such a way that they will be associated with specific *meanings*, and for this
purpose the ontology must provide definitions, conceived as statements of the necessary and sufficient conditions which an entity must satisfy if the term, in its intended
meaning, is to apply to that entity. To say that being an A is a *necessary condition* for
being a B is another way of saying that every B is an A; to say that being an A is a *sufficient condition* for being a B is to say that every A is a B.

A definition, now, is a statement of a set of necessary conditions that are also jointly
sufficient, as in the following example:

X is a triangle = def. *X* is a closed figure; *X* has exactly three sides; each of *X*'s sides is
straight; *X* lies in a plane

Not every statement of necessary and jointly sufficient conditions is a definition.

• The stated conditions used to define the term A must themselves use terms which are
easier to understand (and logically simpler) than the term A itself. (We return to this
issue in our discussion of the "avoid circularity" principle to follow.)

• The stated conditions must be jointly satisfiable. Thus we cannot define, for example, a perpetual motion machine as a prime number that is divisible by 4, even
though everything that is a perpetual motion machine is also a prime number divisible
by 4, and everything that is a prime number that is divisible for 4 is also a perpetual
motion machine (because neither of these things exists or, arguably, could possibly
exist).

14. Use Aristotelian definitions.

The recommended best practice for creating definitions along the lines described earlier is to use the Aristotelian form:

S = def. a G that Ds,

where "G" (for: *genus*) is the immediate parent term of "S" (for: *species*) in the ontology for which the definition is being created. "D" stands for *differentia*, which is to say: "D" tells us what it is about certain Gs in virtue of which they are Ss. Ideally, the terms used in formulating the differentia D will themselves be terms taken from some ontology, where they will themselves be defined.

Consider, as a first example, Aristotle's own definition of "human":

human = def. an animal that is rational

Following this Aristotelian definitional structure ensures that the set of definitions in an ontology precisely mirrors the hierarchy of greater and lesser generality among its universals.

Some examples of Aristotelian definitions from the FMA are as follows:

cell = def. an anatomical structure that has as its boundary the external surface of a maximally connected plasma membrane

plasma membrane = def. a cell component that has as its parts a maximal phospholipids bilayer and two or more types of protein embedded in the bilayer

heart = def. an organ with cavitated organ parts, which is continuous with the systemic and pulmonary arterial and venous trees

liver = def. a lobular organ that has as its parts lobules connected to the biliary tree

lobular organ = def. a parenchymatous organ the stroma of which subdivides the parenchyma into lobes, segments, lobules, and acini[14]

Note that these and other definitions in the FMA ontology contain technical terms such as "cavitated organ part" and "venous tree" that are themselves defined at the appropriate places elsewhere in the ontology.

The Aristotelian definitional structure represents a basic format for the formulation of definitions that can be used regardless of ontological domain, and that is inherently directed at representing the position of each defined term within the relevant *is_a* hierarchy. Addressing the task of formulating definitions can thereby provide an extra check on the correctness of the ontology's *is_a* hierarchy, while creating the *is_a* hierarchy provides an easy first step in the formulation of each definition. These are advantages of the Aristotelian definitional structure, and reasons why it should be adhered to as closely as possible when constructing domain ontologies. Other advantages include:

• Every definition, when unpacked (see below for an explanation of this term), takes us back to the root node of the ontology to which it belongs.

• Circularity is avoided automatically.

• The definition author always knows where to start when formulating a definition.

• It is easier to coordinate the work of multiple definition authors.

Aristotelian definitions work well for common nouns (and thus for the names of universals by which ontologies are principally populated). They do not work at all for those common nouns that are in the root position in an ontology, for here there is no parent term (no genus) to serve as starting point for definition. Root nodes in an ontology must therefore either be defined using as genus some more general term taken from some higher-level ontology, or they must be declared as primitive. Primitive terms cannot be defined, but their meanings can be elucidated (by means of illustrative examples, statements of recommended usage, and axioms—discussed in chapter 7).

15. Use essential features in defining terms.

The definition of a term captures what we can think of as the essential features of the entities that are instances of the designated type. The essential features of a thing are those features without which the thing would not be the type of thing that it is. They are also what we can think of as the constant elements in the structure of the entities in question—those elements that all instances of the relevant universal must possess.

Essential features of a triangle include being a plane figure and having three sides. Inessential features include the lengths of the lines making up the three sides.

For natural objects such as those studied by chemistry, biology, and physics, the essential features of a thing are typically features that play a prominent role in scientific explanation of its existence and behavior. Thus a definition of "portion of water" in terms of essential features would be as follows:

portion of water = def. portion of molecular substance each constituent molecule of which consists of two hydrogen atoms and one oxygen atom connected by covalent bonds

For artifacts, objects deliberately created (or, in some cases, selected) by human beings to be used to achieve certain goals, the essential features have to do with the purpose or use for which the artifact was created. Thus a knife is a tool for cutting things. Its essential features will thus include: having a blade made of a sufficiently hard substance; the blade's having a sharp edge; its having a handle made of some hard substance; its being small and light enough to be manipulated by a single person, and so forth.

What sorts of features will be essential to the objects in a given scientific domain is specified in the relevant scientific literature, and what literature is relevant is determined in turn through specification of the ontology's scope. As can be seen from the definitions provided in our preceding list of examples, essential features of anatomical entities in the FMA include their location in the body, the sorts of anatomical entities that they have as parts, their spatial and physical relationships to other anatomical entities, and so on.

A useful way to go about determining the essential features of entities instantiating a given universal is the following. In the light of available scientific knowledge, attempt to imagine the subtraction and variation of the features of a typical entity falling under the corresponding term, checking at each stage to see whether the considered variation or subtraction would bring it about that the entity in question would no longer be an instance of the universal in question. If such variation or removal of a feature of the entity in imagination has this consequence, then it is highly likely that that feature is one of the essential features of entities of the given kind. Thus it is possible to imagine chairs with different sizes, shapes, materials, and colors; however, the second that one imagines a thing on which it is impossible for a normal human being to sit—for example, that it is made of a soft Jell-O-like material—then it is clear that, whatever the entity being imagined is, it is not a chair, and so the feature of being a thing upon which a normal human being can sit is at least a necessary condition for something's being a chair. Solidity is in this sense a constant (essential) feature of a chair; color a variable feature.[15]

A final point is that, with regard to defined classes, the essential features are just the features mentioned in the definition. Thus the features essential to being a member of the class of people in the United States who have cancer are just to be a person, to have cancer, and to be in the United States.

To see what goes wrong when definition authors fail to utilize essential features of the things being defined, consider again an example from HL7:

person = def. a living subject representing a single human being who is uniquely identifiable through one or more legal documents[16]

16. Start with the most general terms in your domain.

Another recommendation is to define the terms in an ontology from the top down. Thus, in accordance with our Aristotelian template for definitions, we start the process of definition by defining the most general universals first and then working downward through the *is_a* hierarchy toward progressively more specific terms. Beginning with the root, terms on the next level down can be defined by determining relevant differentia in each case. This procedure can be reiterated as many times, and at as many different levels, as are necessary to address identified needs, but allowing the ontology author to start out from the most general level helps to keep things simple at the beginning, and provides a robust perspective from which to address the task of creating a comprehensive ontology at successively deeper levels.

A more general consideration in favor of the top-down approach derives from the proposition that an ontology should have a well-defined and clearly delimited domain, one that is determined, as far as possible, by some preexisting scientific discipline or unified practical field. Beginning with the most general types of entities determined by the specific target domain and working downward from there helps to rule out from the beginning the inclusion in the ontology of content that is not relevant to the chosen domain.

17. Avoid circularity in defining terms.

A definition is circular if the term to be defined, or a near synonym of that term, occurs in the definition itself, as for example in the following:

hydrogen = def. anything having the same atomic composition as hydrogen

poodle = def. anything having the biological structure and physical appearance characteristic of poodles

These definitions are circular because they provide no more information about the nature of the things the terms refer to than do these terms themselves. Since definitions are intended to explain the meaning of a term to someone who does not already understand it, using the term itself or some very similar expression in its own definition defeats the purpose of providing a definition in the first place. Figure 4.2 is a screen shot of the FOAF (Friend of a Friend) Vocabulary Specification 0.99, whose definition of "document" clearly exhibits circularity.

Avoiding circularity is important also for reasons having to do with the correct structuring of an ontology. If we think of a well-built ontology as having the structure of a graph, with a central backbone formed by the taxonomy of the ontology, then for reasons we shall present further on under the heading of "asserted single inheritance," it is important that every node in the graph is linked to the root by means of a unique chain of *is_a* relations. Avoiding circular definitions is a discipline that helps to ensure that this condition is satisfied.

```
Class: foaf:Document

Document - A document.
Status:         stable
Properties      topic primaryTopic sha1
include:
Used with:      workInfoHomepage workplaceHomepage page accountServiceHomepage openid tipjar schoolHomepage
                publications isPrimaryTopicOf interest homepage weblog
Has             Image PersonalProfileDocument
Subclass
Disjoint        Project Organization
With:

The Document class represents those things which are, broadly conceived, 'documents'.

The Image class is a sub-class of Document, since all images are documents.

We do not (currently) distinguish precisely between physical and electronic documents, or between copies of a work
and the abstraction those copies embody. The relationship between documents and their byte-stream representation
needs clarification (see sha1 for related issues).
```

Figure 4.2
Circularity in the FOAF Vocabulary Specification 0.99

18. To ensure the intelligibility of definitions, use simpler terms than the term you are defining.

The terms used in a definition should be more intelligible—for example, by being more scientifically, logically, or ontologically basic—than the term that is being defined. This is to promote the definitions' utility to human beings, for example, to support collaboration across disciplines, by allowing experts in one discipline to use the ontologies prepared for other disciplines as an initial means of orientation. Definitions are used in such cases in order to explain to people who do not know the meaning of a term what that meaning is. Someone who does not know the meaning of a term, especially a technical term, will not be helped by a definition that fails to satisfy the principle of intelligibility.

The following examples, again from HL7, will suffice to illustrate the problem we have in mind:

stopping a medication = def. change of state in the record of a Substance Administration Act from Active to Aborted

health chart entity = def. a health chart included to serve as a document receiving entity in the management of medical records[17]

In scientific contexts—and in the sorts of complex administrative contexts with which HL7 is concerned—it is inevitable that definitions will involve a certain degree of specialized terminology. However if this terminology is to be managed in an effective way, then it is indispensable that for each new step in the direction of greater complexity and of specialization, the terms required are defined using terms already defined in earlier steps, potentially by means of definitional resources imported from other, external ontologies.

19. Do not create terms for universals through logical combination.

From the ontological realist's perspective, that a specific universal exists is never a matter of what can be inferred by logical means alone; it is always only something that must be discovered, through observation and application of the scientific method. Ontology is not analogous to set theory. It embraces what philosophers have referred to as a "sparse" theory of universals, which does not accept that the realm of universals is subject to any rule that allows arbitrary (for example) Boolean combinations.[18]

Thus from the fact that "u" names a universal, we cannot infer that non-u is a universal also, where "non-u" is defined in terms of logical negation as follows:

(*) x instantiates non-u = def. it is not the case that (x instantiates u)

Similarly from the fact that "u" and "v" name universals, we could not infer that "u and v" or "u or v" name universals, defined, respectively, by

(**) x instantiates u and v = def. x instantiates u and x instantiates v

(***) x instantiates u or v = def. either x instantiates u or x instantiates v

We recommend in particular that when building ontologies negative terms should be avoided entirely. That is, the ontology builder should assume that the universals are in every case positive, and so terms such as "nonrabbit" or "nonheart"—defined in accordance with (*)—should not be used, since there are no corresponding negative universals.

This principle applies not merely to terms representing universals, but also to terms representing defined classes. In relation to defined classes, we can formulate the rule as follows:

Avoid postulating complements of classes as entries in an ontology.

The complement of a class is the class containing all of the entities that do not belong in that class. Thus the complement of the class denoted by "dog" is the class denoted by "nondog," a class that includes not merely all cats, all fish, all rabbits, and so forth, but also all cardinal numbers, all musical instruments, all planetary bodies, and everything else that is not a dog.

There are, however, some cases where classes involving a negative element in their definition will properly be included in an ontology.[19] Thus, for example, prokaryotic cells are distinguished from eukaryotic and all other cells precisely by the fact that they lack a cell nucleus. This is, in effect, negative information used to define a class. However, the class of prokaryotic cells is not a complement class. If it were, then it would be equivalent to the class of noneukaryotic *things*, and would thus include, again, all musical instruments, all cardinal numbers, all planets, and so forth. Rather, "prokaryotic cell" is the name of a distinct class of cells that can be clearly defined. It is just that the definition of these cells itself includes some negative information (that they are cells that do not have a nucleus). Only the former formulation ("noneukaryotic thing") is a case of logical or "external" negation. In the latter ("prokaryotic cell"), we have rather internal negation, which from the realist point of view is perfectly acceptable for use in definitions.

The recommendation to avoid negative terms thus needs to be applied with care, since clinical research involves multiple sorts of defined classes referred to by terms in which prefixes like "non-" are used, but which are not defined along the lines of (*). An example is "nonsmoker," which is found in influential health-related terminologies such as SNOMED-CT and MedDRA, and used in scientific assertions such as "nonsmokers are less susceptible to cardiopulmonary diseases than are smokers."

The term "nonsmoker" is perfectly admissible provided it is defined, for example, along the lines of

nonsmoker = def. a human being who does not smoke,

which has exactly the Aristotelian form we recommended earlier.[20] Other putatively negative terms—such as "odorless," "colorless," "invisible," "unfriendly"—are similarly admissible, since they, too, can be defined in a positive way in terms of "lacks".

Another class of negative terms that should be avoided involves the use of negation operators that modify the associated phrases not logically but rather in some other way. Examples are

• canceled oophorectomy
• absent nipple
• unlocalized ligand

In each of these cases we are dealing not with special kinds of entity—there are, strictly speaking, no such things as canceled oophorectomies—but rather with special kinds of knowledge. When we use a term like "canceled oophorectomy" we are talking in an abbreviated way about the fact that "oophorectomy" had earlier been entered into a surgeon's schedule and later removed. Ontologically misleading abbreviations of this sort should not be used when formulating terms in an ontology.

20. Definitions should be unpackable.

Definitions should be substitutable for their defined terms without a change in meaning. If we define an A as "a B that Cs," then we should be able to replace every occurrence of "an A" in a sentence with "a B that Cs," and the result will have the same meaning (and thus also the same truth value) as the sentence with which we began. This process of substitution is called "unpacking."

Note that the unpackability criterion holds only in what are called "extensional" contexts, which means contexts not governed by, for example, expressions such as "John believes," or "In the dictionary it is stated that." In all extensional contexts a defined term should be intersubstitutable with its definition in such a way that the result is grammatically correct and preserves both meaning and truth. The basic idea behind this principle is that, whatever a term refers to (in our case always a universal or defined class), the definition of the term should successfully refer to the very same thing. Thus in the FMA the reference of "heart" should be identical with the reference of the expression "organ with cavitated organ parts, which is continuous with the systemic and pulmonary arterial and venous trees."

The requirement that term and definition be intersubstitutable without affecting meaning is important not only for preserving truth across inference in automated reasoning contexts but also for ensuring intelligibility for human users and maintainers of ontologies. But interchangeability without effect on grammatical correctness is important for human beings also. If replacing a term with its definition results in a grammatically incorrect expression, this will impede the degree to which humans will successfully be able to use the ontology, and it will increase the likelihood of errors.

Principles for Taxonomies

Having set forth principles governing the formulation of definitions, we now move on to principles relating to the role of taxonomies within ontologies.

21. Structure every ontology around a backbone *is_a* hierarchy.

Each ontology should incorporate an *is_a* hierarchy having the structure of a directed acyclical graph with a single root. The terms in the ontology form the nodes of the graph, and the edges represent the *is_a* relation connecting each child to its immediate parent. (In mathematical terms the graph is a directed rooted tree.) The leaf nodes

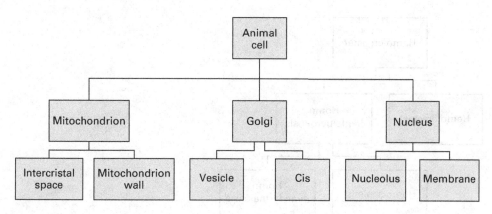

Figure 4.3
Fragment of a partonomy for *animal cell*

(lowest nodes of the ontology) represent the most specific universals or defined classes dealt with by the ontology in its current version. Leaf nodes play no special role in the ontology—since what is a leaf node today may no longer be a leaf node tomorrow because further subtypes have been incorporated.

In addition to those edges representing *is_a* relations, further edges in the graph represent other relations, for example, the *part_of* relationship, which generates what we might call a *partonomy* (as in figure 4.3). (We will discuss the *part_of* relation in more detail in chapter 7.)

Similarly, the relationship *derives_from* can be used to generate hierarchical structures among biological species, as in the simple phylogenetic tree illustrated in figure 4.4.

22. Ensure *is_a* completeness.

The ontology builder should ensure that every term in the ontology is included in its backbone *is_a* hierarchy, and that the ontology is *is_a* complete in the sense that every term in the hierarchy is joined to the root of the tree by a path constituted by successive edges in the graph. Thus if terms are added to the ontology to represent the component parts of the entities for which terms have already been included, then it should be checked that the ontology contains the parent terms needed also for these parts. We note that this principle stands in a mutually supportive relation with the requirement that all terms have definitions constructed using the Aristotelian template (see principle 14), for if this requirement is satisfied then *is_a* completeness will be guaranteed. On the other hand if *is_a* completeness is satisfied, then the creation of Aristotelian definitions is itself more straightforward.

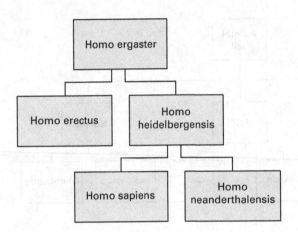

Figure 4.4
A phylogenetic hierarchy

Part of the process of ensuring *is_a* completeness is one of ensuring ontological agreement between terms and their parents. This is achieved by testing the validity of each assertion "A *is_a* B" in a given ontology by checking that, in the relevant domain, every instance of A is an instance of B. This check is needed whether "A" and "B" refer to universals or to defined classes. Bad practice in terminologies often involves the mixing of ontological categories across *is_a* relations, as, for example, in cases such as the following:

nonsmoker is_a finding of tobacco-smoking behavior (from SNOMED-CT);[21]

and, from the still-current version of the Gramene plant environment ontology:

virus is_a plant environment ontology,

unknown environment is_a plant environment ontology,

study type is_a plant environment ontology,

and so forth.[22] Such bad practice would be avoided by application of this simple rule.

23. Ensure asserted single inheritance.

We can think of assertions of the form "A R B," where "A" and "B" are nodes in an ontology and R is a relation that holds between them, as the ontology's *axioms* (for examples, see chapter 7). The ontology builder asserts these axioms during the construction of the ontology. When all the axioms have been asserted, however, then an ontology reasoner such as RACER or FACT (see chapter 8) may *infer* certain further statements. This allows us to distinguish between two different sorts of releases of ontologies: *asserted* and *inferred*.

Our *principle of asserted single inheritance* requires that the central backbone taxonomy of the ontology should be built as an asserted monohierarchy, which means: a hierarchy in which each term has at most one parent.

To speak of "inheritance," here, is to assert that everything that holds of a universal or defined class in an *is_a* hierarchy holds also—*is inherited by*—everything that appears below it in the ontology's *is_a* hierarchy. Because *cat* stands below (is a subclass of) *mammal*, it follows that cats are *vertebrate*, *air-breathing* animals whose females are characterized by the possession of mammary glands. In a similar way, everything that holds of *cell* holds also of *eukaryotic cell*, and everything that holds of *eukaryotic cell* holds also of *plant cell*.

There are a number of reasons for requiring single inheritance in our asserted *is_a* hierarchy. First, adherence to this principle brings certain computational performance benefits.[23] Second, because it ensures that all terms are connected by exactly one chain of parent-child relations to the corresponding root node of the asserted ontology, it provides an easy recipe for the creation of the sorts of definitions we will need in order to apply the Aristotelian template when defining our terms. Indeed, single inheritance is indispensable if the Aristotelian rule is to be applied successfully, since this rule works only if each (nonroot) term in the ontology has exactly one parent.

Third, adherence to single inheritance allows the total ontology structure to be managed more effectively, because it forces the ontology builder to think about each term before positioning it into the ontology, in order to ensure that it is being classified in conformity with the way its neighboring terms are classified. Our own experience with domain experts who are not ontologists and are building ontologies in a variety of different contexts has taught us repeatedly that, when scientists find it difficult to select between multiple parents for a term needing to be included in an ontology, the discipline imposed by the single inheritance principle is welcomed because it repeatedly leads to greater clarity of thinking on the part of those involved.

Fourth, adherence to the principle will make it easier to combine ontologies into larger structures—especially where ontologies need to be combined together automatically.

And finally, and most important, any benefits from multiple inheritance ontologies deriving from their easier surveyability (so that it is easier for human beings to find the terms they need by tracing through multiple parent paths) can be gained in any case by formulating the official (or "asserted") version of an ontology as an asserted monohierarchy and allowing the development of inferred polyhierarchies for specific groups of users. The application ontologies that then result are thus not required to satisfy the principle of asserted single inheritance.

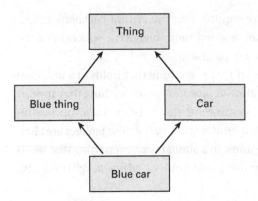

Figure 4.5
A simple illustration of multiple inheritance

Consider figure 4.5, which reflects the attempt to classify things using two different principles of classification (via color and via type of vehicle being classified). The figure does not satisfy the rule according to which (apart from the root) *every node within a hierarchy has exactly one parent.*

Here the principle of single inheritance is violated because two quite distinct *is_a* hierarchies have been run together—a hierarchy of things on the one hand, and a hierarchy of colors on the other. If we need to create such a combined hierarchy for application purposes (for instance to meet the needs of an automobile paint shop), then this can be achieved simply by combining the two mentioned hierarchies together and, using a reasoner, creating an *inferred* hierarchy that will depart from single inheritance but meet our application needs.

To see how resolving apparent multiple inheritance in a classification can be instructive and result in clarification of the correct definitions of the entities involved, consider an assertion such as "some human beings are mothers." What we mean by this assertion is that some human beings at some stage or stages in their lives play the role of mother in relation to other human beings. It may, now, be tempting to classify *mother role* as an example of multiple inheritance, as in figure 4.6.

A treatment along these lines seems to have the advantage that it would allow the variety of different sorts of mother role to be captured in the ontology. At the same time, however, it would gloss over some important distinctions. Recall what the *is_a* relation (represented here by arrows pointing upward) really means:

mother role *is_a* social role ≡ every instance of mother role is an instance of social role

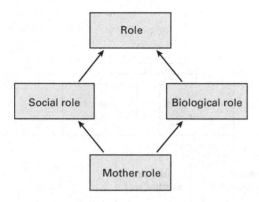

Figure 4.6
The mother role as a putative example of multiple inheritance

mother role *is_a* biological role ≡ every instance of mother role is an instance of biological role

But neither of these assertions is correct, unfortunately. Though often identified by the single word "mother," the social and biological senses of this term are not as closely connected as we might think. It is possible to be a biological mother without playing the social role of being a mother (for example, if one gives one's child up for adoption) and it is possible to play the social role of mother without being a biological mother (if one has adopted a child). Either of the two classifications in figure 4.7 comes closer to a correct way of thinking about the universal *mother* and how it should be classified; and, importantly, neither involves multiple inheritance.

Clearly, it is often possible to classify given individuals in more than one way. For example, pediatric surgeons might be classified both according to their patients and according to the procedures they perform. However, in such cases, the answer is not to create a single taxonomy with multiple inheritance. Rather, one begins by constructing separate "normalized" classifications each using only single inheritance and each built by downward population from a shared upper-level ontology. On this basis, one can then use the definitions of the terms that appear in these two ontologies to spell out the relations between the terms appearing in each. Computer reasoners can then use these definitions to create a compound ontology, in which single inheritance no longer holds, to address specific application purposes.[24]

24. Both developers and users of an ontology should respect the open-world assumption.

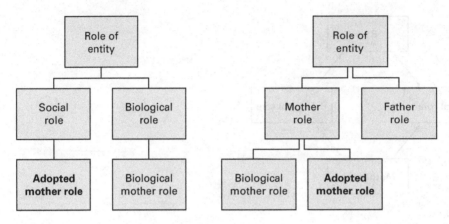

Figure 4.7
Mother classified without multiple inheritance

On the one hand, an ontology should be as complete as possible, given the specific purpose for which it is created. This means: representations of universals should be included in the ontology whenever they are relevant to the purposes of the ontology and fall within its scope. An ideal classification, of the sort that ontologies created for support of scientific research might seek to achieve, would include all existing domain universals at each level in the ontology's *is_a* hierarchy. Thus it would include, for example, all the universals that are discussed in current literature pertaining to the relevant domain. In nontrivial domains such as biology and medicine, of course, this ideal will never be achieved. This is because new scientific information will need to be accommodated with every scientific advance. In such domains an ontology will never be complete, and its authors should make this clear to potential users. Ontologies in general are built on the basis of the *open-world assumption*, or OWA, which means (here) that each ontology is built in a flexible manner to allow extension and correction, and is never put forward as providing a complete assay of the domain in question. Where a new phenomenon is encountered that is relevant to the ontology and falls within its scope, but for which no appropriate term is provided, the proper strategy is to identify the most-specific term in the ontology under which the phenomenon falls, and then to propose an appropriate child term to be added to the ontology hierarchy. The open-world assumption implies that no logical consequences follow from the fact that a given term is *not* included in an ontology.

25. Adhere to the rule of objectivity, which means: describe what exists in reality, not what is known about what exists in reality.

Ontologies created using our recommended methodology are representations of what exists in reality. What exists is not a function of our current state of biological knowledge. The universals treated by natural science in any given domain are discovered, not invented or created.

This is why it is necessary when building an ontology to take into account the best available scientific information about the reality in the salient domain. The goal is to systematically organize the terminological content of that information, paying attention to the essential characteristics of each type of entity. The current state of scientific knowledge is thus crucial for the building of ontologies. Yet the terms included in the ontology *do not refer to our current state of knowledge*. Rather they refer to the corresponding entities in reality. Thus an ontology should not contain classes like: "known allergy," "empirically confirmed boson," or "unclassified influenza." The ontology should not confuse *disease* with *diagnosis* and it should not confuse *result of measurement* with *magnitude that is being measured*.

Conclusion

The process of regimenting the domain information that an ontology contains involves the following steps. First, select the terms that are to be included in the ontology based on domain information that has already been gathered, and distinguish preferred labels and synonyms. Second, provide clear, scientifically accurate and logically coherent definitions for each of these terms. Third, explicitly recognize the location of each of these terms in a hierarchical classification of the domain information. These steps are to be carried out in accordance with the principles listed throughout the chapter. In the next chapter, we will show how these principles are applied in the context of Basic Formal Ontology.

Further Readings on Definitions and Categorization

Ceusters, Werner, and Barry Smith. "A Realism-Based Approach to the Evolution of Biomedical Ontologies." In *Proceedings of the AMIA Symposium*, 121–125. Washington, DC: AMIA, 2006.

Köhler, Jacob, Katherine Munn, Alexander Ruegg, Andre Skusa, and Barry Smith. "Quality Control for Terms and Definitions in Ontologies and Taxonomies." *BMC Bioinformatics* 7 (2006): 212.

Schober, Daniel, Barry Smith, Suzanna E. Lewis, Waclaw Kusnierczyk, Jane Lomax, Chris Mungall, Philippe Rocca-Serra, and Susanna-Assunta Sansone. "Survey-Based Naming Conventions for Use in Ontology Development." *BMC Bioinformatics* 10, no. 125 (2009).

Smith, Barry. "New Desiderata for Biomedical Ontologies." In *Applied Ontology: An Introduction*, ed. Katherine Munn and Barry Smith, 84–107. Frankfurt: Ontos Verlag, 2008.

Smith, Barry, and Werner Ceusters. "HL7 Rim: An Incoherent Standard." *Studies in Health Technology and Informatics* 124 (2006): 133–138.

Smith, Barry, Waclaw Kusnierczyk, Daniel Schober, and Werner Ceusters. "Towards a Reference Terminology for Ontology Research and Development in the Biomedical Domain." In *Proceedings of the 2nd International Workshop on Formal Biomedical Knowledge Representation* (KR-MED 2006), vol. 222, ed. Olivier Bodenreider, 57–66. Baltimore, MD: KR-MED Publications, 2006. Accessed December 17, 2014. http://www.informatik.uni-trier.de/~ley/db/conf/krmed/krmed2006.html.

Examples of Critical Reviews of Ontologies

Bodenreider, Olivier. "Circular Hierarchical Relationships in the UMLS: Etiology, Diagnosis, Treatment, Complications and Prevention." *Proceedings of the American Medical Informatics Association Symposium* 23 (2001): 57–61.

Ceusters, Werner, Barry Smith, and Louis Goldberg. "A Terminological and Ontological Analysis of the NCI Thesaurus." *Methods of Information in Medicine* 44 (2005): 498–507.

Ceusters, Werner, Barry Smith, Anand Kumar, and C. Dhaen. "Mistakes in Medical Ontologies: Where Do They Come From and How Can They Be Detected?" *Studies in Health Technology and Informatics* 102 (2004): 145–164.

Kumar, Anand, and Barry Smith. "The Unified Medical Language System and the Gene Ontology: Some Critical Reflections." *KI 2003: Advances in Artificial Intelligence* 2821 (2003): 135–148.

5 Introduction to Basic Formal Ontology I: Continuants

In previous chapters we have discussed how making use of a formal (or top-level or domain-neutral) ontology can be helpful in constructing domain ontologies that are interoperable, rigorous, and clear. We argued that issues such as terminology selection, term definition, and classification can all be better addressed in the context of a top-level ontology, and also suggested that use of a top-level ontology brings benefits when it comes to sharing ontology content, governance of ontology development, and developing expertise.

Multiple ontologies working together within the framework of the OBO Foundry initiative utilize Basic Formal Ontology (BFO) as a starting point for the categorization of entities and relationships in their respective domains of research. This includes inter alia the Cell Ontology, the Foundational Model of Anatomy, the Protein Ontology, the Ontology for General Medical Science, and the Ontology for Biomedical Investigations. Like other top-level ontologies—and in harmony with the principle of fallibilism—BFO has been and is still subject to ongoing review and to multiple different sorts of testing by the developers and users of these and many other ontologies supporting empirical science. BFO has been modified, and we hope improved, as a result of the lessons learned by the many groups who have been applying the ontology to particular domains, and also in reflection of considerable input from external critics. In this chapter we will introduce the categories of BFO along with their definitions, focusing on version 2.0 of the ontology that was released for comment in 2014. We will also describe how these categories can be applied to an existing domain ontology by what might be thought of as a process of reverse engineering.

Some Basic Features of BFO

BFO is an upper-level ontology developed to support integration of data obtained through scientific research. It is deliberately designed to be very small, in order that it should be able to represent in consistent fashion those upper-level categories common to domain ontologies developed by scientists in different fields. It is small, too, in order to allow exercise of the benefits of modularity and of the division of expertise; a top-level ontology should not contain terms like "cell," "death," or "plant" that properly belong in domain-specific ontology modules of narrower scope.

Because of its generality and small size, BFO will not in and of itself address the terminological needs of those working in specific domains; it provides rather a starting point for work by those with specialist knowledge. Nor will BFO provide complete answers to many of the quasi-philosophical questions that arise in the course of domain-specific ontology development—questions such as "what is an organization?" or "what is life?" or "what is a work of art?" or "what is an action?" This does not mean that answers to such questions cannot be expressed within the BFO framework. In some cases—as we shall see in our discussion of information artifacts to follow—such debates, involving both BFO developers and users, have led to the recognition of the need to extend BFO itself, though still without departing from the narrow confines of what is domain neutral.

BFO assists domain ontologists by providing a common top-level structure to support the interoperability of the multiple domain ontologies created in its terms. In this way it helps to bring about a situation in which information compiled in separate repositories can form part of a common framework for the categorization of, and for reasoning about, the entities in the corresponding domains. This will be achieved, of course, only if informaticians, data managers, researchers, and curators of experimental literature and data, do in fact utilize BFO in their work. At the same time, however, the adoption of BFO has been shown to bring benefits that provide additional justification for its use. Thus, for example, its use promotes portability of expertise, so that those who have once been trained in use of BFO in one domain can more easily apply their skills in other areas. It also provides a starting point for ontology development, which allows those new to ontology to move more quickly to the sorts of domain-specific questions which belong to their respective areas of expertise by providing a set of ready-made answers to the abstract questions with which top-level ontology is concerned. Thus many people use BFO-based ontologies without being aware that they are doing so; the BFO components of these ontologies remain invisible to them in their work, in much the same way that the engine of a car is most of the time unnoticed by the driver.

In previous chapters we have discussed the distinction between universals and particulars, and we have stressed that the primary goal of scientific ontologies is the representation of universals and the relations between them. Ontologies, like scientific theories, are concerned with capturing knowledge of what is general. The terms in BFO, and in the ontologies developed on its basis, should thus be understood, in the first place, as representing universals. This is not because of an interest in universals for their own sake. The ultimate goal of scientific ontologies is to support the work of scientists in *classifying particulars*, for example, the sorts of particulars they observe in their experiments. BFO supports the construction of classificatory hierarchies to aid in reasoning about such particulars. It provides a set of preconfigured high-level taxonomic distinctions that can serve as an off-ramp for the population of representations of universals at lower levels of generality. BFO is thus designed to help ensure that the domain ontologies built on its basis represent the universals in their respective domains in a consistent and coherently structured fashion.

Basic Types of Entity: Continuant and Occurrent

BFO takes as its starting point an interest in the workings of reality from the point of view of those who are engaged in scientific research. We take reality to be comprised of entities—using "entity" in a sense that is common among both philosophers and scientific researchers to refer to anything at all that exists in any way at all.

We then divide entities into two categories, namely:

• *Continuants*: entities that *continue* or *persist* through time, including (1) independent objects (for example, things such as you and me); (2) dependent continuants, including qualities (such as your temperature and my height), and functions (such as the function of this switch to turn on this light); together with (3) the spatial regions these entities occupy at any given time (see figure 5.1);

• *Occurrents*: entities that *occur or happen*, variously referred to as "events" or "processes" or "happenings," which we take to comprise not only (1) the processes that unfold in successive phases but also (2) the boundaries or thresholds at the beginnings or ends of such processes, as well as (3) the temporal and spatiotemporal regions in which these processes occur.

These two types of entities do not exist side by side with each other in any simple sense. Rather, in keeping with the doctrine of perspectivalism outlined in chapter 3, they correspond to two distinct and complementary perspectives on one and the same reality, neither of which can do full justice to those features of reality represented by the other.

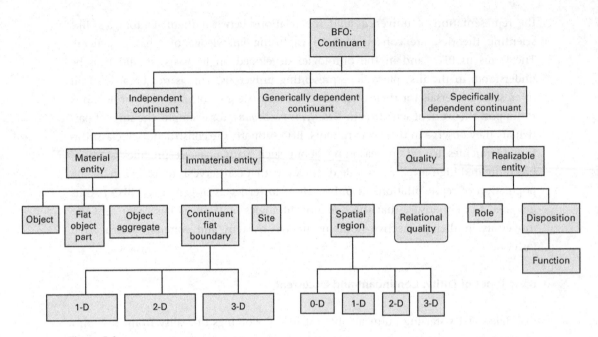

Figure 5.1
The hierarchy of BFO continuants

In describing this reality, we customarily draw on both of these perspectives simultaneously, as when we make assertions such as

• there are cells (continuants) engaging in processes of meiosis (occurrents);

• there are people (continuants) having surgeries (occurrents) performed on them by
other people (continuants);

• there are amino acid chains (continuants) that participate in processes of folding
(occurrents), which result in new structures (continuants) that themselves participate
in processes of posttranslational modification (occurrents) resulting in typical three-
dimensional amino acid chain structures (continuants);

• there is the earth (continuant) that orbits (occurrent) the sun (continuant).

Note how, in each such case, we talk about particulars, such as particular things and
particular processes by using general terms such as "amino acid chain" or "orbit" referring to universals.

In what follows, we will describe the *is_a* hierarchy of BFO types and subtypes,
beginning with the top-level categories and working downward through the subtypes,

first for *continuants* and then, in chapter 6, for *occurrents*. We will proceed in order of diminishing concreteness, beginning with independent continuants and material entities.

BFO: Continuant

Continuants in BFO are entities that *continue to exist through time*; they may gain and lose parts (for instance, as an organism gains and loses cells), but at each point in time at which they exist at all they nonetheless exist completely. Thus you may lose an arm, but you—all of you, at whatever time you exist—will still exist completely, and this is so in times both before and after the loss of the arm. The loss of an arm may be more painful than, but it is ontologically comparable to, the loss of a single hair. This is in contrast to *processes* (which form the main subcategory of occurrents), which unfold themselves through time in successive temporal parts or phases. Because no two distinct phases exist simultaneously, there is no point in time at which a process exists as a whole. Rather, it exists at any given point in time only in some correspondingly short-lived stage or slice.

The continuant portion of BFO consists of representations of entities that (1) persist, endure, or continue to exist through time while maintaining their identity, and (2) have no temporal parts. Note that this is not a *definition* of "continuant." This term is so basic to our understanding of reality that it is not possible to provide a definition that does not itself use terms, like "persist" or "endure," which are equivalent in meaning. Any attempted definition will thus be circular. We provide, instead, what we can think of as an elucidation of what we mean by the term "continuant," together with examples designed to illustrate the sorts of entities to which the term is to be applied. (And where "BFO:" terms are introduced in what follows in the absence of a definition, this is because a similar policy is being applied in these cases as well.)

While continuants do not have temporal parts, each continuant will be associated with a life or, in BFO parlance, a *history*—an entity belonging to the realm of occurrents, and thus the sort of thing that can have temporal parts.

Examples of continuants include a tomato, the qualities of the tomato—for example, its weight or temperature or color, and the region of space occupied by a tomato at any given time. BFO's *continuant* has subtypes intended to capture all of these types of continuant. Its three immediate subtypes or "children" are: *independent continuant, specifically dependent continuant,* and *generically dependent continuant.*

BFO: Independent Continuant

An *independent continuant* is a *continuant* entity that is the bearer of qualities. If a continuant entity *a* is the bearer of quality *b*, then we also say that *b inheres in a*. Thus the color *b* of tomato *a inheres in* tomato *a*. Inherence itself can be defined as a kind of one-sided dependence, more precisely as that sort of one-sided dependence that obtains among qualities, dispositions, and roles (an explanation follows) on the one hand and independent continuants on the other. There are other uses of the word "dependence," some of which will concern us in what follows (for instance in our treatment of the relation between a boundary and that which it bounds). For the moment, however, it is crucial to understand the very specific sense of "dependence" upon which BFO relies, a sense of dependence that implies that the dependent entity is secondary (has diminished concreteness) in relation to the independent continuant that is its bearer. The latter is a three-dimensional thing that has material parts. The dependent entity, by contrast, has no material parts but is rather parasitic on the material thing that supports it. Material things cannot be parasitic on (or ontologically secondary to) other entities in this sense. (There is nothing more concrete than material things.) And from this it follows that an independent continuant, while it is an entity in which other entities (such as qualities) inhere, cannot itself inhere in anything.

Independent continuants are such that their identity and existence can be maintained through gain and loss of parts, and also, as we shall see, through changes in their qualities, and through gain and loss of dispositions, and of roles. Tomato *a* may be left out in the sun and lose its moisture; tomato *a* may once have been green, but is now red; tomato *a* may be frozen, and thereby lose its disposition to ripen; and tomato *a* may be selected by the chef, and thereby acquire the role of garnish to your steak.

Types of independent continuants to be dealt with in what follows include organisms and their parts—for example, your heart and the collection of your limbs; the *boundaries* of organisms—for example, your fingertips (the sorts of things used to take impressions called "fingerprints"); and *places*, such as the Grand Canyon. BFO correspondingly distinguishes two subtypes of independent continuant: *material entity* and *immaterial entity*.

BFO: Material Entity

A BFO: *material entity* is an independent continuant that has some portion of matter as part. It is thus an independent continuant that is spatially extended in three dimensions, and that continues to exist through some interval of time, however short.

Examples of BFO: *material entity* are organisms such as human beings, undetached arms of human beings, and aggregates of human beings such as, for example, a dance troupe or a baseball team. These three sets of examples correspond to the three principal subtypes of material entity distinguished by BFO, namely:

- BFO: *object*
- BFO: *fiat object part*
- BFO: *object aggregate*

together with various combinations that will be discussed.

BFO: Object

Nature is organized as a hierarchy of nested units. From microphysical particles to planetary bodies, there are units or grains in the order of reality—referred to in what follows as "objects." Examples are atoms, molecules, organelles, cells, organs, organisms, planets, and stars.

An *object* is a material entity that is

1. spatially extended in three dimensions;

2. causally unified, meaning its parts are tied together by relations of connection in such a way that if one part of the object is moved in space then its other parts will likely be moved also (the parts share in this sense what we can think of as a *common fate*);

3. maximally self-connected (which means intuitively that the different parts of the object are tied together in a certain way and that anything that is tied to these parts in the same way is itself part of the object).[1]

An organism is an object in this sense, as is a single cell, an egg (including all its contents), a space ship (including all its contents), and a planet. Two people shaking hands, in contrast, do not form an object, nor does one person joined together with his hat. This is because the connections are in both cases too weak to join the parts together in the sense required. Your (attached) head does not form an object even though the connections between its parts are physically sufficiently strong for objecthood. This is because your head, unlike your body as a whole, is not maximal in the sense required.

In many cases, an object enjoys the requisite sort of self-connectedness in virtue of possessing a physical covering layer or membrane, a container that holds the parts inside it together. The covering layer may have holes or cavities but these are in normal circumstances too small to allow the objects contained within it to escape. The covering layer is itself both topologically self-connected and maximal. It is self-connected in the sense that, selecting any portion of the layer, we can trace a continuous (though not always straight, given that the surface may include holes) line to any other portion of the layer without needing to go outside it. And it is maximal in the sense that every portion of matter to which we can trace a similarly continuous line is included as part of the layer.

A tomato is an object in this sense, not however the two halves of a tomato before separation. A human being, too, is an object. And as both of these examples show, the fact that the surface of an object must be self-connected does not imply that the surface does not contain holes—for example, pores, or your mouth—through which particles of matter can penetrate in one or other direction. An organ such as your heart or brain is an object in this sense, and so also is a fetus. Each of these entities is connected by physical conduits to its surrounding host organism. But these connections are relatively weak, and (as we know from experience) the object in question is able to survive its disconnection.

The possession of a maximally self-connected outer boundary—called an *object boundary* in the BFO ontology—works well as a criterion of objecthood for macroscopic objects, which means, roughly, for independent continuants at least as large as a cell nucleus. However, it cannot be applied to serve as such a criterion for objects at finer grains—roughly, of single molecules or smaller. At such levels the criterion of causal unity plays a more central role. This criterion is applied both internally, where it relates to the ways in which the different parts of the object are related to each other, and externally, where it has to do with the ways in which the object as a whole interacts causally with other objects.

Even where an object has a maximally connected physical *outer* boundary, it may still include in its interior parts that are not connected to its other parts. An example is provided by the blood cells in your body. These are parts of your body, though they are not connected to the other parts.[2] The bacteria that form your microbiome are located in the interior of your body, but they are neither connected to your body nor a part thereof.[3] They do, however, in virtue of the surrounding membrane that is your skin, share a common fate.

Objects are the bedrock upon which dependent continuants and occurrents depend for their existence. An object is an entity that can exist and be what it is regardless of what other objects exist. Thus, a doorknob is an object because it can be removed from a door and still exist with all its parts intact. It can be moved from one place to another, and survive even when the objects around it are destroyed, removed, or replaced. Organisms, which as we saw are objects in the BFO sense, can certainly be said to depend on other objects (for example, on oxygen, water, drugs, food for sustenance, and so forth); these senses of "dependence" are not however of significance for us here.

We said that BFO continuants *exist in full at any time in which they exist at all*. Again: if Jill loses her arm, then she still exists as a whole—with all the parts that she currently has—and if her arm survives this loss, it is now a *separate object*. It is no longer a part of Jill. The reason for insisting on this point is that something similar is not true in the case of BFO's *occurrents*. The first set of a tennis match may now be in the past, but it is still part of the whole tennis match, and will indeed always remain so, just as the first three years of your life are still and will always remain a part of your life. Lives, in this sense, and processes in general, have temporal parts. Continuants do not have temporal parts.

Not only do the relations between an occurrent and its parts hold atemporally, but so also do the relations between an occurrent and its properties. Occurrents, as we shall see, behave differently, from a logical point of view, from the ways continuants behave.

BFO: Object Aggregate

An *object aggregate* is a *material entity* that is made up of a collection of objects and whose parts are exactly exhausted by the objects that form this collection. In addition the objects forming an object aggregate are separate from each other in the sense that they share no parts in common.

Examples of object aggregates are a heap of stones, a group of commuters on the subway, a population of bacteria in your blood, a flock of geese, the collection of patients in a hospital. The degree of unity of such entities is, we might say, weaker than that possessed by objects proper—compare a heap of stones to a single stone. In some types of object aggregates, the objects themselves may interact dynamically, as, for example, in the case of a symphony orchestra, or an infantry battalion going into battle.

Organizations such as symphony orchestras or tenants' associations are object aggregates of a special sort, in which specific objects (specific human beings), play specific roles (for example, president, secretary, treasurer, member, and so on). (See our discussion of roles that follows.)

BFO: Fiat Object Part

A *fiat object part* is a *material entity* that is a proper part of some larger object, but is not demarcated from the remainder of this object by any physical discontinuities (thus it is

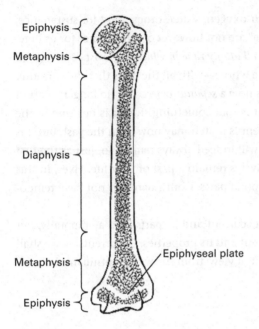

Epiphysis
Metaphysis
Diaphysis
Metaphysis
Epiphyseal plate
Epiphysis

Figure 5.2
A long bone such as a femur divided into three different types of fiat parts: *epiphysis*, *diaphysis* and *metaphysis*
Source: From http://medical-dictionary.thefreedictionary.com/epiphysis, originally published in Miller-Keane and Marie T. O'Toole, *Miller-Keane Encyclopedia and Dictionary of Medicine, Nursing, and Allied Health*, 7th ed. (London: Elsevier, 2003).

not itself an object). Examples of fiat object parts include your upper torso, the handle of a solid metal spoon, the Western hemisphere of the planet earth, the diaphysis of a long bone (see figure 5.2)

Fiat object parts are to be contrasted with those object parts that are objects in their own right, and which thus have complete physical outer boundaries of their own, for example, the blood cells in your veins and arteries or the individual sardines in a can of sardines. We use the term "fiat" to draw attention to the fact that the boundaries in question are standardly the reflection of decisions on the part of, for example, the drawer of a map or the theorist identifying regions of different sorts within a domain of continuous variation. Such divisions are drawn even where there are no physical discontinuities to which the dividing lines correspond. As we shall see in our discussion of fiat object boundaries to follow, some boundaries of this sort exist even in the absence of any decision by a cognitive agent.

Combination Object-Entities

BFO makes no claim to the effect that objects, object aggregates, and fiat object parts provide an exhaustive classification of the types of material entity. Thus, for example, if John owns two neighboring apartment buildings, but sells the top floor of one of them, then it may be that what he owns is the sum of an object together with a fiat object part. If Mary is studying knee injuries in a population of patients then it may be that the target of her study is an aggregate of fiat object parts.

Combination entities of this sort[4] provide no special challenges to BFO. They are not included explicitly as subtypes in BFO 2.0, though they could be included in future versions of the ontology if there is a corresponding need on the part of BFO's users. We may distinguish further subtypes of fiat entities analogously also in the realm of two-dimensional surfaces (for example, Arizona is a fiat part of the two-dimensional surface that is the continental United States, which may in turn be a fiat part of the two-dimensional surface of the planet earth).

BFO: Specifically Dependent Continuant

In line with our strategy of moving through BFO's *is_a* hierarchy in order of diminishing concreteness, we deal next what BFO calls *specifically* and *generically dependent continuants*. A *specifically dependent continuant* is a *continuant* entity that depends on one or more specific independent continuants for its existence. Dependent continuants exhibit *existential dependence* in the sense that, in order for a dependent continuant to exist, some other entity in which it inheres (intuitively, an entity enjoying a larger degree of concreteness) must exist also.

Examples of specifically dependent continuants include the color of this tomato, the pain in your left elbow, the mass of this cloud, the smell of this piece of mozzarella, the disposition of this fish to decay, your role of being a doctor, the function of your heart to pump blood, and the quality of a specific pixel array on your screen. The mass of this cloud could not exist without this cloud and the color of this tomato could not exist without this tomato.

In BFO, *specifically dependent continuants* are subclassified as follows:

BFO: *quality*

    BFO: *relational quality*

BFO: *realizable entity*

    BFO: *role*

BFO: *disposition*

BFO: *function*

A *specifically dependent continuant* is a *dependent continuant* that depends on some specific independent continuant that is its bearer. Thus a specifically dependent continuant is such that it cannot migrate from one bearer to another. My suntan is specifically dependent on me. It cannot also be *your* suntan, however closely similar the two distinct instances of the suntan type might be. Similarly, the mass of your car exists only so long as the car exists, and that very instance of *mass* can only exist as the mass of this specific car, and not of some other car. As we shall see, this is not true of generically dependent continuants, which are defined by the fact that they can migrate from one bearer to another.

BFO: Quality

There are two types of *specifically dependent continuant*: *quality* and *realizable entity*. Qualities are contrasted with realizables in that the former, if they inhere in an entity at all, are fully exhibited or manifested or realized in that entity. The latter, in contrast, can inhere without being realized, and can be realized to different degrees (including different degrees of likelihood).

What all qualities have in common is that they inhere in, and so depend on, other entities; in order for a quality to exist some other entity or entities—specifically, one or more independent continuants, must also exist. Examples of qualities include the mass of this kidney, the color of this portion of blood, and the shape of this hand. Notice that, in each of these cases, the quality is referred to as standing in a relationship to some other independent continuant entity, such as a kidney, a portion of blood, a hand. This is because of the dependent nature of qualities. There cannot be color without it being the color *of something*, and there cannot be mass without it being the mass *of something*. In particular, it is BFO independent continuants that qualities depend on (or as we also say: inhere in). Qualities may inhere in one independent continuant—for example, the shape quality of this glass cube; or they may inhere in multiple independent continuants—for example, the quality of *being siblings* or *being competitive with* that might inhere in John and Mary; the quality of being angry that might inhere in an aggregate of persons we call a mob.

Qualities may depend also on entities of other types; thus, for example, the quality of your heart, of *beating with a certain rate*, is dependent not only on your heart, but also on the beating process in which the heart participates.

Often, it is the qualities of objects and their parts that we refer to as differentiae when formulating definitions. Why is our sun classified as a star? Because it shares certain qualities with other celestial bodies already identified as stars, such as being self-luminous, being plasmatic, having a size, mass, and temperature within a certain range, and so on.

Qualities in BFO can be ordered into those more or less general, for example, as follows:

quality

  color

    red

      dark red

        RGB 990033 dark red

or

    body temperature

      elevated body temperature

        body temperature in the range 37.6° and 38° Celsius

          body temperature of 37.8° Celsius

Note that to assert that a body temperature quality of 37.8° Celsius inheres in a certain body does not imply that the body has a temperature of 37.8° Celsius uniformly through all its parts. Rather, different temperature qualities will be detectable in different parts of the body, and the "37.8° Celsius" represents an average of these.

More general qualities are called "determinables" in the philosophical literature, and the corresponding least general qualities are called "determinates." Typically, determinable qualities—for example, *mass*—hold of independent continuants as a matter of necessity. (You cannot, as a matter of necessity, exist without your mass.) Determinate qualities—for example, *mass of 70 kg*—hold only contingently. A red nose need not be red, but it must have some color. We can think of the necessary or essential determinable qualities as constants within the architecture of their bearers, while the determinate qualities vary with time. You always have your temperature, but the value of your temperature varies from one time to the next.

BFO: Relational Quality

Relational qualities have a plurality of independent continuants as their bearers. Examples include a marriage bond, an instance of love, being a parent of, and so on, all of which obtain between one person and another. From the BFO perspective there is both the relational-quality universal *marriage bond* (an entity that might be included in a domain ontology for social reality) as well as the specific instances of this universal obtaining between (and so specifically depending upon) John and Mary, Bill and Sally, and so forth.

Relations That Do and Relations That Do Not Have Instances

By contrast with BFO's *relational qualities*, relations such as **instance_of** or **part_of** (discussed in more detail in chapter 7), are relations for which it does not make sense to speak of instances. They are not entities in their own right. If it is true that Mary is a human being then there is no extra entity—for example, no instance of the relation of instantiation—that is needed to make this true. If it is true that Mary's heart is **part_of** Mary, then similarly there is no extra entity in addition to Mary and her heart that is needed to make it true that this relation obtains. BFO adopts this view of relations for reasons of practical utility. (We have, for instance, no data pertaining to different instances of the parthood relation.)

Internal relations such as comparatives (*taller-than, larger-than, heavier-than . . .*) are also not entities in their own right, as BFO conceives them.[5] If John is taller than Mary, then this is accounted for exclusively in terms of John's and Mary's respective height qualities, and in terms of the fact that each of these heights instantiates a certain determinate height universal and that the totality of such universals form a certain linear order. (And we note that "fact" here is not being used as a technical term, and thus also not as referring to an extra entity in the BFO sense.)

BFO: Realizable Entity

Like qualities, *realizable entities* are *specifically dependent continuants* that inhere in one or more independent continuants. Realizable entities are *in* (inhere in) their bearers in just the same way that qualities are in their bearers. In contrast to qualities, however, realizable entities are exhibited only through certain characteristic processes of realization.

A *realizable entity* is thus defined as a specifically dependent continuant that has at least one independent continuant entity as its bearer, and whose instances can be realized (manifested, actualized, executed) in associated processes of specific correlated types in which the bearer participates.

Examples of *realizable entity* include the role of being a doctor, the functions of the reproductive organs, the disposition of a portion of blood to coagulate, the disposition of a portion of metal to conduct electricity. Entities in each of these types are in each case associated with entities of corresponding process types in which they are realized (executed, manifested, actualized). Thus, for example, the role of a doctor is realized when he examines or treats patients; the function of a reproductive organ is realized in copulation or insemination.

Realizable entities are entities of a type whose instances are such that in the course of their existence they contain periods of actualization, when they are manifested through processes in which their bearers participate. But they may also exhibit periods of dormancy, when they exist by inhering in their bearers but without being manifested—as, for example, in the case of those diseases which are marked by periods of latency, and by the many occupational roles that are not realized because the bearer is, for example, asleep. Some realizable entities are realized during all the times when the bearer exists, as, for example, in the continuous functioning of a mammal's heart and lungs; other realizable entities are realized hardly at all, as, for example, in the case of sperm, which are nevertheless the bearers of a function—to carry male genes to a female's egg—that are entities *of a type* some (an important few) of whose instances do indeed contain periods of actualization.

We saw that an animal is classified as a mammal in virtue of possessing certain qualities (being a vertebrate, being warm blooded); but there are characteristics of mammals that involve not qualities but realizables. For example, a (female) mammal is capable of giving birth to live young, and capable of lactating, even if in particular cases these dispositions are never realized. Some realizables, as again in the case of the function of a sperm to penetrate an ovum, may be such that they can be manifested only once in their lifetime. Others, for example, the function of a spark plug in an internal combustion engine, can be manifested over and over again.

BFO distinguishes two principal types of *realizable entity*: *role* and *disposition*, with one subtype of *disposition*, namely *function*. Other subtypes of realizable entity distinguished in the ontological literature include *capabilities* and *tendencies*.[6]

BFO: Role

A *role* is an externally grounded *realizable entity*, that is, it is a realizable entity that is possessed by its bearer because of some external circumstances (for example, the bearer has been assigned the role by some other persons, who have roles of their own which grant them a certain authority). A role is thus always optional; the bearer does not have to be in the given external circumstances.

Each instance of *role* is a realizable entity that (1) exists because the bearer is in some special physical, social, or institutional set of circumstances in which the bearer does not have to be (optionality), and (2) is not such that, if this realizable entity ceases to exist, then the physical make-up of the bearer is thereby changed (external grounding).

A role is a realizable entity whose manifestation brings about some result or end. This result is not essential to its bearer in virtue of the kind of thing that it is. Thus it is not essential to Jim that he is a nurse, or to Mary that she is a bankrupt or a baroness or a bodyguard. But Jim and Mary have these roles because they are in certain associated kinds of natural, social, or institutional contexts. A role can cease to exist without the physical make-up of the bearer thereby being changed. An entity has a role not because of the way it itself is, but because of something that happens or obtains externally.

Further examples include the role of an instance of a chemical compound to serve as analyte in an experiment, the role of a portion of penicillin to serve as a drug, and the role of a stone in marking a boundary.

As we shall see, roles are distinct from another type of realizable entity, called functions. A heart has the function of pumping blood, but in certain circumstances that same heart can play the role of dinner for a lion or of plastinated prop in a museum display. A portion of water does not have any function per se, but it does play many different roles, for example, in a hydroelectric experiment, or in washing clothes, or in helping to initiate the growth process of a seed. Many prominent types of role involve social ascription. A person can play the role of lawyer or of surrogate to a patient, but it is not necessary for persons that they be lawyers or surrogates.

Note, here, that it is the role that is ontologically prior. "Lawyer role," "surrogate role," "nurse role," and so on, all refer to universals in BFO's terms. "Lawyer," "surrogate," "nurse," however, refer merely to defined classes. If a person has the role of lawyer, then we can refer to the person in two ways, as person, or as lawyer. The latter usage can be defined as follows:

$lawyer(x)$ = def. $person(x)$ and for some $y(lawyer\text{-}role(y)$ and x **has_role** $y)$

This definition can serve as a template for very many role-related defined classes.

When once the class lawyer has been defined, then it may be used in BFO-based ontologies in many of the same ways these ontologies use terms representing universals, thus, for example, in assertions of the form: lawyer is_a person.

BFO: *roles* are specifically dependent instances. A role exists only when some specific independent continuant serves as its bearer. Roles in this sense, like qualities, cannot migrate from one bearer to another. The term "role" can, however, be used in a different sense in contexts such as Jane's being the seventh person to fill the role of director of this institute, or Joe's being the third person to play a particular role in a play. "Role" in this sense is being used to designate what BFO calls a *generically* dependent continuant.

BFO: Disposition

It is common for researchers to make claims such as

- element X has a disposition to decay into element Y,
- the cell wall is disposed to filter chemicals in endocitosis and exocitosis,
- certain people have a disposition to develop colon cancer,
- children are innately disposed to categorize objects in certain ways.

All of these are examples of dispositions in BFO's sense. A *disposition* is a *realizable entity* in virtue of which—for example, through appropriate triggers—a process of a certain kind occurs (or can occur or is likely to occur) in the independent continuant in which the disposition inheres. This process is called the *realization* of the disposition. The trigger might consist in the objects being placed in a certain environment or being subjected to certain external influence, or it may be some internal event within the object itself.

Unlike a role, a *disposition* is a realizable entity that is such that, if it ceases to exist, then its bearer is physically changed. Dispositions are in this sense (and in contrast to what is the case with roles) not optional. If an entity is physically a certain way, then it has a certain disposition, and if it ceases to be that way, then it loses that disposition. A disposition can thus be conceived of as an *internally grounded* realizable entity. That is, it is a realizable entity that exists because of certain features of the physical make-up of the independent continuant that is its bearer. One can think of the latter as the material basis of the disposition in question. Note that this material basis will exist even though its associated disposition is never realized.

Dispositions are variable along a continuum from weaker to stronger. Dispositions at the weaker end of the spectrum are not realized in every suitable triggering situation, but only in some fraction of relevant cases. Examples include

- a hemophiliac's disposition to bleed an abnormally large amount of blood, and
- the disposition of a person who smokes two packs of cigarettes a day throughout adulthood to die of a disease at a below average age.

Clearly, we are often referring to more or less weak forms of disposition when we consider genetic and other risk factors for specific diseases.

By contrast, we can distinguish a strong form of disposition, a *sure-fire disposition*, which is reliably executed whenever its bearer is in the conditions appropriate for a disposition of the corresponding type. Examples include

- the disposition of a piece of stretched elastic to contract when released,
- the disposition of a sheet of glass to break if struck with a sledgehammer moving at 100 feet per second,
- the disposition of a diploid cell to become haploid following meiosis, and
- the disposition of a magnet to attract iron filings.

Incorporation of dispositions into the BFO ontology provides a means to deal with those aspects of reality that involve possibility or potentiality without the need for complicated appeals to modal logics or possible worlds. At the same time, the ontological commitment to dispositions itself faces the problem as to how dispositions are to be individuated. If John has the disposition to scratch his nose, does he also have the disposition to scratch his nose *when awake*, or *in the presence of Mary*, or *during a full moon*? How, in other words, are dispositions to be counted? How is one disposition to be distinguished from another? BFO's approach to answering such questions is highly practical. BFO has been created to serve the annotation of data deriving from scientific experiments. BFO itself does not provide a taxonomy of dispositions; it does not itself legislate concerning which types of dispositions exist, or how they are to be individuated. Rather, it leaves this task to the specific sciences. Those involved in scientific practice have at their disposal at each stage a limited repertory of terms for representing the salient types of dispositions, and it is this set of evolving repertories that will serve as starting point for ontology building in the spirit of BFO. Scientific practice does not reduce the massive diversity in the number of ways in which the totality of dispositions can be divided up, nor does it solve all problems concerning how dispositions are identified or individuated; but it does solve the practical problem of providing us with a means to represent those dispositions in each given domain that are salient to scientific advance.

BFO: Function

A *function* is a special kind of disposition.[7] It is a realizable entity whose realization is an end-directed activity of its bearer that occurs because this bearer is (a) of a specific kind and (b) in the kind or kinds of contexts that it is made or selected for. Thus a function is a disposition that exists in virtue of the bearer's physical make-up, and this physical make-up is something the bearer possesses because of how it came into being—either through natural selection (in the case of biological entities) or through intentional design (in the case of artifacts).[8] Roughly, the entities in question came into being in order to perform activities of a certain sort, called "functionings." Examples include

- the function of amylase in saliva to break down starch into sugar,
- the function of a sperm to penetrate an ovum,
- the function of a hammer to drive in nails,
- the function of a pen to write, and
- the function of a heart pacemaker to regulate the beating of a heart by means of electricity.

Each function has a bearer with a specific type of physical make-up. This is something that, in the biological case, the bearer has evolved to have (as in a hypothalamus secreting hormones) and, in the artifact case, something that the bearer has been designed and built to have (as an Erlenmeyer flask is designed to hold liquid).

It is not accidental or arbitrary that the eye has the function to see or that a screwdriver has the function of fastening screws. Rather, eyes and screwdrivers exist because they perform these functions. Their functions are integral to the entities in question in virtue of the fact that the latter have evolved, or been constructed, to have the physical make-up needed to perform or realize them. It is because of its physical make-up that your heart's function is to pump blood and not, for example, to produce thumping sounds—the latter are mere byproducts of your heart's functioning.

Like dispositions in general, therefore, functions are internally grounded realizable entities: a function is such that if it ceases to exist, then its bearer is physically changed. If a lung or attic fan becomes nonfunctioning, then this indicates that the physical makeup of these things has changed. In the case of the lung this might be due to a cancerous lesion; in the case of the attic fan to a rusted exhaust screen.[9]

BFO: Specifically Dependent Continuant: Summary

Examples of the different kinds of specifically dependent continuant recognized by BFO are

- this negative charge is a *quality* of this phosphate ion,
- this adhesion is a *quality* of the water in this flask,
- John's obligation to pay Susan is a *relational quality* that obtains between John and Susan,

- to detoxify its containing organism is a *function* of this liver,
- to produce portions of glycogen is a *function* of this endoplasmic reticulum,
- this bacterium in this case of cholera has the *role* of pathogen,
- this person in this clinical trial has the *role* of subject,
- this rattlesnake has the *disposition* to strike when threatened, and
- this structure of mature bamboo scaffolding has the *disposition* to be cyclone-resistant.

The BFO ontology of dispositions serves as the basis for the treatment of diseases in the BFO-based Ontology for General Medical Science.[10] To say that a human being has a case of influenza, for example, is to say that he or she has a complex disposition that is realized, inter alia, in acute inflammation, weakness, dizziness, and fever. A person may also have a *pre*disposition to some disease without in fact having the disease. Many persons, for example, have a predisposition to colon cancer; we may have this predisposition for the whole of our lives without ever developing the disease of colon cancer itself. In this case we have a disposition (already now) to acquire a further disposition at a later time. In a similar way, each healthy adult human being has a disposition to walk. A human fetus has a predisposition to walk; that is, she has a disposition to acquire the disposition to walk at a later stage in her life.

Reciprocal Dependence among Realizable Dependent Continuants

Consider the cases of husband and wife, or of doctor and patient. Here pairs of reciprocally dependent roles are involved, whereby the first role in each reciprocal pair can be realized only if the second is realized also. We encounter analogous reciprocally dependent pairs of functions in the realm of artifacts. Consider a key and the associated lock. The key has a disposition to unlock the lock, while the lock itself has the disposition to be unlocked by that very key. Both dispositions are manifested in the same process, namely, in the key's unlocking of the lock. What underlies these complementary dispositions is the key's disposition to transmit torque when rotated, the lock's disposition to release when unlatched, and a relation between the material and shape qualities of the lock and key that confers these dispositions (the key must fit the lock and must be of sufficient hardness to enable transmission of torque to the lock's lever).

Reciprocally dependent pairs of functions are present throughout the natural world. Consider the case of sperm and egg. Here biological functions have evolved in complementary dependence upon each other. Each cannot realize its primary function unless the other does so also.[11]

BFO: Generically Dependent Continuant

To say that one entity is *specifically dependent on* another is to assert that the first entity is as a matter of necessity such that it cannot exist unless the second entity exists. BFO's specifically dependent continuants are thus subject to what we might call the axiom of nonmigration: they cannot migrate from one bearer to another. Some dependent continuants seem, however, to be capable of such migration, as, for example, when you copy a pdf file from one computer to another. Clearly the pdf file is *dependent* on some bearer; for the pdf file to exist, there must be some physical storage device on which it has been saved. But equally clearly, the pdf file can be *moved* from one storage device to another. The very same pdf file can be saved to multiple storage devices, and thus it—the numerically identical information entity—can exist in multiple copies.

To do justice to this and many similar phenomena BFO incorporates the category of *generically dependent continuant*, defined as a continuant that is dependent on one or other independent continuants that can serve as its bearer. More formally we define *generic dependence* as follows:

a generically depends on *b* = def. *a* exists and *b* exists and: for some type *B*, *b* is an instance of *B* and necessarily (if *a* exists then some *B* exists)

and we define generically dependent continuant on this basis:

a is a generically dependent continuant = def. there is some *b* such that *a* is generically dependent on *b*

If *A* is a subtype of generically dependent continuant, then every instance of *A* requires some instance of independent continuant subtype *B*, whereby which instance (or instances) of *B* serves as bearer can change from time to time.

There are two large families of examples of such entities—in the domains of information artifacts and of biological sequences respectively. And while BFO itself does not contain terms like "information artifact" or "DNA sequence," terms like these are found in the Information Artifact Ontology (IAO)[12] and Sequence Ontology (SO),[13] both of which are BFO conformant.

We can think of generically dependent continuants, intuitively, as complex continuant patterns of the sort created by authors or designers, or (in the case of DNA sequences) through the processes of evolution. Generically dependent continuants thus include, for example, the Coca Cola trademark, the pattern that is your signature, a square arrangement of sixty-four alternating black and white squares. Each such pattern exists only if it is *concretized* in some counterpart specifically dependent continuant—the pattern of red and white swirls on the label of this Coca Cola bottle; the pattern of ink marks you just created by signing this piece of paper; the pattern of black-and-white squares on this chessboard.

Such patterns can be highly complex. The pattern of letters of the alphabet and associated spacing which is the novel *Robinson Crusoe* is concretized in the patterns of ink marks in this (and that) particular *copy* of the novel. Generically dependent continuants can be concretized in multiple ways; you may concretize a novel in your head.[14] You may concretize a piece of software by installing it in your computer. You may concretize a cake recipe that you find in a cookbook by reading it, and your concretization may then serve as the starting point for a process of creating a plan, which exists as a realizable dependent continuant in your head and is realized in your baking of a cake.

Generically dependent continuants may be found in the realm of nucleic acid and other biological sequences. Other generically dependent continuants are information entities created by human beings. The data in your database, for example, are patterns in some medium—for instance in your hard drive—with a certain kind of provenance. The database itself is an aggregate of such patterns. When you create the database you create a particular entity (what BFO calls an "instance") of the generically dependent continuant type *database*. This will be concretized in your hard drive as a certain complex quality (of magnetic excitation)—a specifically dependent continuant. Similarly each entry in the database is an instance of the generically dependent continuant type *datum*, which will be concretized in your hard drive as a certain part-quality of that whole quality that is the concretization of the database as a whole.

Databases, novels, dramatic scripts, musical scores, and other information entities are in some ways analogous to other created artifacts such as paintings or sculptures. They differ from the latter in that, once having been created, they can exist in many copies that are all of equal value. The novel *Robinson Crusoe* is an instance of the type *novel*, each printed copy is an instance of the type *book*. The novel *Robinson Crusoe* is a generically dependent continuant instance, an *abstract pattern*, made concrete through the acts involved in printing successive copies. In each of these copies there inheres a certain complex quality (of a certain quantity of bound paper and associated small piles of printer's ink), and each such complex quality concretizes the generically dependent continuant that is Defoe's novel.

In this way we can do justice to the fact that there is only one *Robinson Crusoe*, which does not change when additional copies are printed.

In the case of a work of music such as Beethoven's *Symphony No. 9,* there is again a certain abstract pattern, a generically dependent continuant instance of the type *symphony*, which is itself a subtype of the type *musical work*, which is concretized in certain specifically dependent patterns of ink marks that we find in a printed copy of the score or in certain specifically dependent patterns of grooves in a vinyl disk. The symphony is *realized* (manifested, performed) in those occurrent patterns of air vibrations that are instances of the type *musical performance*. The score itself is an instance of the generically dependent continuant type *plan specification*, which is concretized in the minds of the conductor and the members of the orchestra when they read and understand the score. This allows them to create (and to realize as they perform) a plan, which is a complex, realizable dependent continuant that exists (in slightly different but mutually compatible forms) in the minds of multiple human beings; it is realized when conductor and orchestra work together to create the already mentioned pattern of air vibrations.

Analogously, when a research team decides to perform an experiment following a published protocol, the protocol itself is a generically dependent continuant instance of the type *plan specification*. The leader of the research team concretizes this protocol as a complex quality in her mind by reading it, and creates on this basis that specifically dependent realizable continuant that is a plan for carrying out this experiment. At the same time she creates a series of subprotocols, plan specifications for her various team members, which are concretized by them as plans for carrying out their corresponding parts of the experiment. The experiment itself is the synchronized realization of these plans.

BFO: Immaterial Entity

Having dealt with BFO's specifically and generically dependent entities, we now return to the other major subclass of BFO's *independent continuant*, namely *immaterial entity*, defined as an independent continuant that contains no material entities as parts. Even to speak of "immaterial entities" may sound, at first, counter-intuitive. However, consideration of cases makes it clear that there are entities in reality that although not themselves material are nonetheless important for our manipulation and cognition of what is material. A good example set of such entities is found in the domain of anatomy, where the boundaries of, for instance, organs and portions of tissue are no less salient than the entities that they bound. Rosse and Mejino provide the following rationale for including terms for immaterial entities such as surfaces, lines, and points in the Foundational Model of Anatomy (FMA) ontology: "Although anatomical texts and medical terminologies with an anatomical content deal only superficially, if at all, with anatomical surfaces, lines, and points, it is nevertheless necessary to represent these entities explicitly and comprehensively in the FMA in order to describe boundary and adjacency relationships of material physical anatomical entities and spaces."[15]

Immaterial entities divide into two major subgroups:

1. *Boundaries* and *sites*, which bound, or are demarcated in relation to, *material entities*, and which can thus change location, shape and size as their material hosts move or change shape or size (for example, your waist, the boundary of Wales [which moves with the rotation of the earth]; your nasal passage, the hold of a ship);

2. *Spatial regions*, which exist independently of *material entities*, and which thus do not change.

Immaterial entities listed under 1. are in some cases **continuant parts** of their material hosts. Thus the hold of a ship, for example, is a part of the ship; the hold may itself have parts, which may have names (used, for example, by ship stow planners, customs inspectors, smugglers, and the like). Immaterial entities under both 1. and 2. can be of zero, one, two, or three dimensions.

Sites, such as the kitchen of a restaurant on a ship, are analogous to material entities in that they are of three dimensions and can move through space. When they do so they will occupy successively different spatial regions. One site may move through another site, for instance the interior of a railway carriage may move through the Mont Blanc tunnel. By contrast, spatial regions never move through each other, because spatial regions never move. (More precisely, they are, by definition, at rest relative to the pertinent frame of reference, as will be discussed.)

BFO: Continuant Fiat Boundary (including Zero-, One-, and Two-Dimensional Continuant Fiat Boundary)

A *continuant fiat boundary* is an *immaterial entity* that is of zero, one, or two dimensions and does not include a *spatial region* as part. Intuitively, a *continuant fiat boundary* is a boundary of some material entity that exists exactly where that object meets its surroundings. For BFO: *objects* larger than molecules, the *fiat object boundary* is its maximally connected two-dimensional surface, for example, the surface of the earth, or the surface of a cell membrane. However, a fiat boundary can also be the boundary of an immaterial entity, such as a site (for example, the boundary of a portion of airspace into which only military aircraft are allowed to fly).

In the simplest cases such as rocks or baseballs, and even in topologically more complex cases such as donuts or wedding rings or bird cages, there is little difficulty in

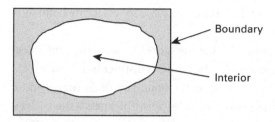

Figure 5.3
A block of marble with corrosive acid eating outward toward its boundary

determining where the corresponding object boundary lies. In the case of compartmentalized objects such as mammals, buildings, and refrigerators, however, we may face options as to whether to include the compartments (cavities) within the object as *parts* of the object or as *holes*. Consider, for example, your digestive tract. On one view your body is topologically analogous to a donut; your digestive tract is a hole running through the middle. On the view espoused in the FMA, however, which is the de facto standard human-anatomy ontology, the digestive tract is not a *hole in* but rather a *part of* the organism—a part that is not made of matter. Similarly the interior of your freezer compartment is not a *hole in* but is rather a *part of* your refrigerator. Whichever option we take will determine what is to be counted as the "outer" boundary of the object in question and thus also of the object's shape.

Note that the sense of "boundary" that is presupposed in the preceding is one according to which boundaries have no material parts. Entities with material parts are in every case spatially of three dimensions. Continuant boundaries as we conceive them are always of lower dimension.

Consider a rectangular block of marble. The surface of the block is a boundary of two dimensions, its edges are of one dimension, and its corners are of zero dimension. Each of these boundaries is dependent on the cuboid, but in a sense of "dependence"—which we can call "boundary dependence"—that is different from the sense employed when dealing with specifically and generically dependent continuants above. Briefly, we can say that a boundary *a* of an object *b* is boundary-dependent on this object if and only if it is necessarily such that it can exist only if *either b* exists *or* there exists some part of *b* that includes *a* as part. To see what is at issue here, imagine that there is some capsule of a supremely powerful corrosive acid inside the marble block that is eating the marble away, by degrees, from the inside (figure 5.3). As the marble is progressively destroyed its boundaries are at first unaffected. They will continue to exist for just as long as there is at least *some* remaining part of the block that includes them as part.

Since this remaining part can be arbitrarily thin, there is a sense in which the boundary itself is of zero thickness.

It will be clear from the preceding discussion that the sense of "boundary" intended here—which is close to the mathematical sense of the term—is distinct from that which is involved when we refer to a skin or membrane as the boundary of an organism or cell. Material boundaries in this latter sense—boundaries with thickness—themselves have boundaries (on either side) of the type at issue here.[16]

Continuant fiat boundaries admit of different dimensions. A *two-dimensional continuant fiat boundary* (surface) is a self-connected fiat surface whose location is defined in relation to some material entity. Examples of this type of boundary include any surface of a continuant material object that segments that object off from the rest of its environment, such as the boundary of the block of marble in the example just discussed. A *one-dimensional continuant fiat boundary* is a continuous fiat line whose location is defined in relation to some material entity; for example, the Greenwich meridian, the Equator, and geopolitical boundaries of nations and states. Finally, a *zero-dimensional continuant fiat boundary* is a fiat point whose location is defined in relation to some material entity. Examples include the North Pole and the point of origin of a spatial coordinate system.

Boundaries and Granularity

Why, now, does BFO refer to object boundaries as *fiat*, given that the outer boundary of, for example, a tomato or a block of marble or a table in our living room does not depend for its existence on any decision or on any drawing of boundaries by any cognitive agent? The answer to this question turns on BFO's treatment of the phenomenon of granularity.

If we examine the surface of the table with a powerful microscope, then it will appear that there is no boundary there at all, in either the *mathematical* or the *thin layer* sense just distinguished. Rather, there is just (something like) a cloud of microparticles oscillating at high velocities in the vicinity of what, when we use the naked eye, we like to call the surface of the table. In a famous passage by the physicist Eddington on what he called "My Two Tables," the view that there are no boundaries of (middle-sized) objects—and so there are no corresponding (middle-sized) objects for them to bound— is defended explicitly. Table 1, as Eddington sees it, is the ordinary solid table made of wood; table 2 is what he called his "scientific table": "My scientific table is mostly emptiness. Sparsely scattered in that emptiness are numerous electric charges rushing about with great speed; but their combined bulk amounts to less than a billionth of the bulk of the table itself. [The scientific table] supports my writing paper as satisfactorily as table No. 1; for when I lay the paper on it the little electric particles with their headlong speed keep on hitting the underside, so that the paper is maintained in shuttlecock fashion at nearly steady level."[17]

Eddington here expresses the sort of reductionist point of view that we rejected in chapter 3 (especially the section on adequatism). For him only the scientific table exists; table 1 is for him something like a convenient fiction. From the adequatist point of view defended by BFO, in contrast, denying that the two tables have just the same degree of reality is a mistake, since the two tables are in fact one and the same object— it is merely that they are viewed at different levels of granularity. Table 1 and table 2 have the same degree of reality in the same way as do, for example, the City of Toronto depicted on a large- and a small-scale map (where the former shows items at the order of magnitude of single streets and houses, the latter only major highways and neighborhood divisions).

The fiat object boundaries of tables and tomatoes exist, because the tables and tomatoes exist, as is seen when these objects are viewed from the perspective of medium-sized-object granularity. These fiat object boundaries are referred to (implicitly or explicitly) when we apply the commonsensical distinction between what is in the interior and what is in the exterior of the objects in question. Something similar occurs also when we use a map to determine what is in the interior and what is in the exterior of some parcel of real estate. This does not mean that those who wish to embrace a reductionist view cannot use BFO to support their work in ontology development. Reductionists who wish to follow Eddington can simply ignore (not use) those parts of BFO pertaining to boundaries at higher levels of granularity. For most users of BFO however, and especially users of BFO in areas of biology and the health sciences, its adequatist framework provides them precisely with the resources they need to deal ontologically with collected data pertaining to boundaries in both of the two distinguished senses. This is clear from the large number of terms for surface boundaries (in addition to surface layers) found in the FMA. Representing surface boundaries is important, too, in areas such as perceptual psychology—for example, in experiments on vision that gather data pertaining to surface colors and to perceptual surfaces of different shapes and textures.

The issue here pertains to distinctions of granularity in scientific research and in clinical practice, in engineering, administration, and other practical disciplines. Different scientific specialties explore the same domains of reality at different levels of granularity, and what are counted as objects on one level of granularity may appear to scientists working on another level of granularity as object aggregates. To describe BFO

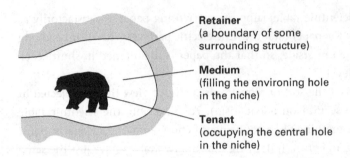

Retainer
(a boundary of some
surrounding structure)

Medium
(filling the environing hole
in the niche)

Tenant
(occupying the central hole
in the niche)

Figure 5.4
A site containing a bear
Source: Barry Smith and Achille Varzi, "Surrounding Space: The Ontology of Organism-
Environment Relations," *Theory in Biosciences* 121 (2002): 139–162.

as an "adequatist" ontology is to say that it is designed to support the work of scientists
and engineers at multiple scales and levels of aggregation, and thereby also to support
the integration of data relating to such multiple levels. Different BFO users may address
the problem of such integration in different ways. Some may be able to ignore this
problem because they focus exclusively on one level of granularity. Others may need to
annotate each of BFO's material entity types with explicit reference to the level of
granularity at issue, and work is ongoing to create an extension of BFO 2.0 in which
resources for such explicit reference will be provided.[18]

BFO: Site

A *site* is, intuitively, an immaterial entity in which objects—such as molecules of air or
water, or a bear—are or can be contained (see figure 5.4).

A site exists because there is some material entity in relation to which it is defined,
providing, for example, the floor and walls and ceiling that allow containment by
forming the *retainer* for the site. Each site will thus have a characteristic spatial shape in
virtue of this physical retainer. But the site itself, while it exists because of this retainer,
does not contain the retainer as part. The site is, rather, the *hole* that is contained by the
retainer. A BFO: *site* can now be defined as follows:

site = def. a three-dimensional *immaterial entity* that either (1) is (partially or wholly)
bounded by a *material entity* or (2) is a three-dimensional immaterial part of an entity
satisfying (1)

Examples include your nasal cavity, your veins (cavities through which blood flows),
the Suez Canal (trench), the lumen of your gastrointestinal tract, the interior of your

aorta, the interior of your office, the trunk of your car, the Piazza San Marco, a kangaroo pouch, the inside of your shoe, your eye socket, the cruciform slot of a Philips head screw.

All of these examples are at the levels of granularity accessible to ordinary human perception. The Protein Ontology Consortium is developing a sub-ontology representing amino acid sites that are the locations of post-translational modifications.[19] We leave open the question whether there are BFO: *sites* at other levels of granularity. Are, for example, black holes BFO: *sites*? Such questions will need to be addressed empirically, in light of the consequences of applying BFO to corresponding domains.

Every site will at any given moment coincide with some spatial region. But the site is not *identical* with the spatial region with which it coincides, because the site but not the spatial region is ontologically (site-) dependent upon its retainer. In the case of mobile sites (for example, a ship's cargo hold) the site in question will cycle through a continuous multiplicity of spatial regions as its retainer, the ship, moves. While at any given moment the hold will be co-extensive with some spatial region, it is not *identical* to that spatial region, because the hold remains what it is even after the ship and the sites for which its interior walls serve as retainers have moved so as to occupy new spatial regions. Spatial regions cannot move since it is spatial regions in and through which movement occurs.

A site is typically associated with some medium such as the body of air that is partially or completely enclosed by its retainer. Thus, the nasal cavity is a site that is formed by the exterior boundaries of the inner membranes and parts of the nose that give this site its characteristic spatial shape. The site serves as container for a succession of molecules of oxygen and nasal flora. Similarly, the skull is a site that contains the cranial cavity, the brain, and the cerebrospinal fluid that entirely fills the cavities that are enclosed by the skull walls and the brain, taken together.

BFO: Spatial Region (including Zero-, One-, Two-, and Three-Dimensional Spatial Regions)

A spatial region is a continuant entity that is a part of space (by which we mean: a part of the maximal or total space, or in other words of the whole of space). Both material and immaterial entities occupy regions of space; processes occur in space. When an object moves from one place to another, then it occupies a continuous series of different three-dimensional spatial regions at different times. As we know from the theory of relativity, however, there are no spatial regions except as defined relative to some frame of reference, an issue we discuss in the next section.

BFO recognizes four different sorts of spatial regions in its ontology, of three, two, one and zero dimensions. Just as, from the ontological realist perspective, there are objects (independent continuants such as you and me) so there are three-dimensional spatial regions that such objects occupy. And just as for BFO there are surfaces of objects (for example, the two-dimensional external fiat boundary of your body) so there are two-dimensional spatial regions that these boundaries occupy and one- and zero-dimensional spatial regions that are the boundaries of these boundaries.

BFO thus incorporates two levels of continuants—with *spatial regions* on one level and *material entities* and *sites* (with their respective boundaries) on the other, the former providing, as it were, the spatial receptacles for the latter. Such a two-level approach is common in the literature on spatial reasoning.[20]

BFO is a boundary-tolerant ontology.[21] It incorporates terms for spatial regions of zero, one, two, and three dimensions (points, lines, surfaces, and volumes, respectively) and also terms for the objects, fiat object boundaries, and sites that occupy the corresponding spatial regions. (As we shall see, BFO adopts a similar two-level theory in its treatment of temporal entities.) One rationale for recognizing the two levels of objects and the regions they occupy turns on the way in which the part-whole structures of objects reflect the part-whole structures of the corresponding spatial regions. In decomposing an object into its constituent parts we also decompose the spatial region occupied by the object into the spatial regions occupied by these parts at any given time. If two parts of the object are connected to each other, then so also are the corresponding spatial regions (and this is true independent of the frame of reference we are using to determine the reference of spatial region identifiers in any given case).

In accepting sites into its ontology in addition to spatial regions, BFO is acknowledging also two distinct location relations involving independent continuants of

1. **containment**, between an independent continuant and a site that contains it (for example, between a chick and the interior of an egg, or between a group of drinkers and the interior of a pub), and

2. **location**, between any independent continuant entity and the corresponding spatial region—whereby every independent continuant is, at any given time, associated with the spatial region at which it is located at that time.

Independent continuants may have many qualities (such as shape, size, mass, density, reflectance, electric charge, and so forth), stand in many different sorts of relations to other entities, and be such that many realizable dependent continuants inhere in them.

Spatial regions, in contrast, are continuants of a peculiar ("abstract") sort. There is a sense in which they have qualities of shape and size, but the primary BFO relation here is one of instantiation between a spatial region instance and the corresponding spatial region universal. Spatial regions can be said to have the quality of being size *m because* they instantiate the universal *spatial region of size m.*[22] The corresponding qualities, accordingly, are "defined qualities," and form a special subfamily of defined classes in the BFO framework.

Spatial regions do not inhere in any other entities; and they are inert, in the sense that no realizables inhere in them. Spatial regions are thus unique kinds of entities in BFO. They are entities in the full sense; however, they are neither material entities of the sort that they provide locations for, nor are they dependent on such concrete material entities in the way that qualities and realizables are.

As we have seen, while some determinable qualities—such as *mass*—hold of independent continuants as a matter of necessity, the corresponding qualities of lower generality such as *mass of 70 kg* hold only contingently. When it comes to qualities and relations of spatial regions, in contrast, all hold as a matter of necessity at all scales. A spatial region cannot change its shape, since if it did then this would mean that it had ceased to exist and had been replaced by some other spatial region. Similarly a spatial region cannot change its relations (for example, of adjacency or parthood) to other spatial relations.

The four subtypes of BFO: spatial region are as follows:

1. BFO: *zero-dimensional region* is a spatial region with no dimensions, also called a spatial point.

2. BFO: *one-dimensional region* is a spatial region with one dimension, also called a spatial line.

3. BFO: *two-dimensional region* is a spatial region with two dimensions, also called a spatial surface.

4. BFO: *three-dimensional region* is a spatial region with three dimensions, also called a spatial volume.

Spatial Regions and Frames of Reference

As pointed out above, spatial regions cannot be specified absolutely but always only relative to some reference frame, and work is ongoing to create a future version of BFO in which such reference frames will be incorporated explicitly. For most current users of BFO, however, the effects of special relativity are not significant and thus they can safely make the assumption that there is a single Euclidean frame of reference that they all—modulo trivial differences, for example, as to choice of origin, or of coordinates used—share in common.

A reference frame is, in first approximation, a system of coordinates with an origin and units. And (to our knowledge) all coordinate systems employed by current users of BFO are easily convertible one into another. This is because significant problems of conversion between coordinate systems arise only where frames of reference are in motion relative to each other. When we are all working with what is effectively the same Newtonian frame because we are dealing with objects and spatial regions close to or on the surface of the earth, such relative motion is insignificant. In some cases a system of coordinates is specified in an experimental protocol—for instance when observations are being made of animal behavior using coordinates defined relative to a specific forest. Users of BFO should document such specifications explicitly when employing BFO: spatial region terms in annotations. In other cases the coordinates are provided by some standard, as for instance in the case of the representation of latitude and longitude on a map. The map then divides up the represented land and sea surfaces into (roughly) rectangular two-dimensional spatial regions, and we can think of the lines themselves as representing one-dimensional spatial regions, and of the points where they intersect as representing zero-dimensional spatial regions.

The spatial regions defined by a reference frame are always at rest relative to this frame. Thus in the cases normally treated by biologists and clinical scientists the spatial regions they refer to (for example, a lab bench or hospital ward) can be assumed to be at rest—they can be treated as if they were absolute containers for the things and events observed—and all space-related measurements, for example, of speed or of relative distance can be directly compared.

In the future we anticipate that BFO will be used in support of domain ontologies developed for many different types of research, some of which may involve frames of reference that are not at rest relative to each other. A space transport ontology, for example, might include a reference frame that, because it is in motion relative to the earth, is not trivially interconvertible with the standard Newtonian frames used by biologists. Such conversions can be made, but may be quite complex—as, for example, where demarcations of spatial regions in terms of the World Geodetic System (WGS-84) need to be converted into demarcations in accordance with the International Celestial Reference System (ICRS) maintained by the International Astronomical Union (IAU).

Where, as in some areas of physics, BFO-based domain ontologies contain representations of spatial regions that are defined in terms of what are called noninvariant frames of reference, a special situation arises, since convertibility here may not be achievable. Future versions of BFO will be required to provide appropriate means to support the development of domain ontologies of this sort, and as we shall see, analogous issues will arise also with regard to BFO categories of temporal region.

A BFO: Continuant Classification

Having outlined the *continuant* perspective of BFO, we conclude by providing a simple illustration of how BFO might be used to provide a classification of the qualities, functions, and dispositions relating to the human heart.

- this human heart **instance of** *object,*
- this heart's surface **instance of** *fiat object boundary,*
- this collection of four hearts in a biobank **instance of** *object aggregate,*
- this superior vena cava **instance of** *fiat object part,*
- this biopsy sample of the septum of the heart **instance of** *material entity,*
- this mediastinum **instance of** *site,*
- this mass of 250 grams **instance of** *quality,*
- this disposition to deteriorate over time **instance of** *disposition,*
- this disposition to pump blood **instance of** *function,*
- this role of serving as plastinated prop **instance of** *role.*

Further Reading on Basic Formal Ontology

Basic Formal Ontology website: http://www.ifomis.org/bfo.

Chapters 5, 6, and parts of 7 are based on the draft specification of BFO 2.0, which contains also formal definitions of the terms introduced in these chapters as well as associated axioms and theorems and considerable further explanatory material. This specification, which can be found at the BFO website, will be updated at intervals in advance of the official release of BFO 2.0.

Grenon, Pierre, and Barry Smith. "SNAP and SPAN: Towards Dynamic Spatial Ontology." *Spatial Cognition and Computation* 1 (2004): 1–10.

Grenon, Pierre, and Barry Smith. "A Formal Theory of Substances, Qualities and Universals." In *Proceedings of the International Conference on Formal Ontology and Information Systems* (FOIS 2004), ed. Achille Varzi and Laure Vieu, 49–59. Amsterdam: IOS Press, 2004.

Smith, Barry, and Pierre Grenon. "The Cornucopia of Formal Ontological Relations." *Dialectica* 58 (2004): 279–296.

Smith, Barry. "On Classifying Material Entities in Basic Formal Ontology." In *Interdisciplinary Ontology (Proceedings of the Third Interdisciplinary Ontology Meeting)*, ed. Barry Smith, Riichiro Mizoguchi, and Sumio Nakagawa, 1–13. Tokyo: Keio University Press, 2012.

Further Reading on Granularity

Rector, Alan, Jeremy Rogers, and Thomas Bittner. "Granularity, Scale and Collectivity: When Size Does and Does Not Matter." *Journal of Biomedical Informatics* 39 (3) (2006): 333–349.

Smith, Barry, and Berit Brogaard. "A Unified Theory of Truth and Reference." *Logique et Analyse* 43 (169–170) (2003): 49–93.

Vogt, Lars. "Spatio-Structural Granularity of Biological Material Entities." *BMC Bioinformatics* 11 (2010): 289.

Further Reading on Independent Continuants

Bittner, Thomas, Maureen Donnelly, and Barry Smith. "Individuals, Universals, Collections: On the Foundational Relations of Ontology." In *Formal Ontology and Information Systems: Proceedings of FOIS 2004*, ed. Achille Varzi and Laure Vieu, 37–48. Amsterdam: IOS Press, 2004.

Casati, Roberto, and Achille Varzi. *Holes and Other Superficialities*. Cambridge, MA: MIT Press, 1994.

Simons, Peter. "Particulars in Particular Clothing: Three Trope Theories of Substance." *Philosophy and Phenomenological Research* 54 (1994): 553–575.

Smith, Barry. "Fiat Objects." *Topoi* 20 (2001): 131–148.

Smith, Barry, and Achille Varzi. "The Niche." *Noûs* 33 (2) (1999): 198–222.

Smith, Barry, and Achille Varzi. "Fiat and Bona Fide Boundaries." *Philosophy and Phenomenological Research* 60 (2000): 401–420.

Smith, Barry, and Achille Varzi. "Surrounding Space: The Ontology of Organism-Environment Relations." *Theory in Biosciences* 121 (2002): 139–162.

Varzi, Achille. "Boundaries, Continuity, and Contact." *Noûs* 31 (1997): 26–58.

Further Reading on Dependent Continuants

Ariew, A., R. Cummins, and M. Perlman, eds. *Functions: New Essays in the Philosophy of Biology and Psychology*. Oxford: Oxford University Press, 2002.

Batchelor, Colin, Janna Hastings, and Christoph Steinbeck. "Ontological Dependence, Dispositions and Institutional Reality in Chemistry." In *Formal Ontology in Information Systems:*

Proceedings of the Sixth International Conference (FOIS 2010), ed. Antony Galton and Riichiro Mizoguchi, 271–284. Amsterdam: IOS Press, 2010.

Bird, A. *Nature's Metaphysics: Laws and Properties*. Oxford: Oxford University Press, 2007.

Dipert, Randall. *Artifacts, Art Works, and Agency*. Philadelphia: Temple University Press, 1993.

Fine, Kit. "Ontological Dependence." *Proceedings of the Aristotelian Society, New Series* 95 (1995): 269–290.

Goldfain, Albert, Barry Smith, and Lindsay G. Cowell. "Dispositions and the Infectious Disease Ontology." In *Formal Ontology and Information Systems (Proceedings of FOIS 2010)*. Amsterdam: IOS Press, 2011.

Jansen, Ludger. "The Ontology of Tendencies and Medical Information Science." *The Monist* 90, Special Issue on Biomedical Ontologies (2007): 534–555.

Martin, C. B. "Dispositions and Conditionals." *Philosophical Quarterly* 44 (1994): 1–8.

Further Reading on Boundaries, Spatial Regions, and Topology

Bittner, Thomas. "A Mereological Theory of Frames of Reference." *International Journal of Artificial Intelligence Tools* 13 (1) (2004): 171–198.

Bittner, Thomas, and Maureen Donnelly. "A Temporal Mereology for Distinguishing between Integral Objects and Portions of Stuff." In *Proceedings of the Twenty-Second AAAI Conference on Artificial Intelligence (AAAI)*, ed. R. Holte and A. Howe, 287–292. London: Elsevier, 2007.

Casati, Roberto, Barry Smith, and Achille Varzi. "Ontological Tools for Geographic Representation." In *Formal Ontology in Information Systems: Proceedings of the First International Conference (FOIS 1998)*, ed. Nicola Guarino, 77–85. Amsterdam: IOS Press, 1998.

Casati, Roberto, and Varzi Varzi. *Parts and Places: The Structures of Spatial Representation*. Cambridge, MA: MIT Press, 1999.

Cohn, A. G., and J. Renz. "Qualitative Spatial Representation and Reasoning." In *Handbook of Knowledge Representation*, ed. F. van Harmelen, V. Lifschitz, and B. Porter, 551–596. Amsterdam: Elsevier, 2008.

Cohn, Anthony G., and Achille Varzi. "Mereotopological Connection." *Journal of Philosophical Logic* 32 (4) (2003): 357–390.

Donnelly, Maureen. "Relative Places." In *Formal Ontology in Information Systems: Proceedings of the Fourth International Conference (FOIS 2004)*, ed. Achille Varzi and Laure Vieu, 249–260. Amsterdam: IOS Press, 2004.

Donnelly, Maureen. "A Formal Theory for Reasoning about Parthood, Connection, and Location." *Artificial Intelligence* 160 (2004): 145–172.

Donnelly, Maureen. "Containment Relations in Anatomical Ontologies." In *Proceedings of the AMIA Symposium*, 206–210. London: Elsevier, 2005.

Galton, Anthony. *Qualitative Spatial Change*. Oxford: Oxford University Press, 2001.

Haemmerli, Marion, and Achille Varzi. "Adding Convexity to Mereotopology." In *Formal Ontology in Information Systems*, ed. Achille Varzi, 65–78. Amsterdam: IOS Press, 2014.

Smith, Barry. "Smith, Barry. "Boundaries: An Essay in Mereotopology." In *The Philosophy of Roderick Chisholm*, ed. Lewis Hahn, 534–561. LaSalle: Open Court, 1997.

Smith, Barry. "Mereotopology: A Theory of Parts and Boundaries." *Data & Knowledge Engineering* 20 (1996): 287–303.

6 Introduction to Basic Formal Ontology II: Occurrents

Having presented the *continuant* categories of BFO, we will now focus on the *occurrent* categories starting, again, with the most general, and then working downward through their respective subtypes. The occurrent portion of BFO represents entities that occur, happen, unfold, or develop in time. In commonsensical terms, these entities are occurrences or happenings or the processes of change; they are the ontological counterparts of present participles (runnings, swimmings, dividings, orbitings).

A BFO: *occurrent* is, more precisely, either an entity that unfolds itself in time, or it is the instantaneous boundary of such an entity (for example, a beginning or an ending) along what we can think of as the time axis, or it is a temporal or spatiotemporal region that such an entity occupies. *Occurrent*, correspondingly, has four subtypes:

BFO: *process*

BFO: *process boundary*

BFO: *temporal region*

  BFO: *spatiotemporal region*

BFO: Process

A BFO: *process* is an *occurrent* entity that exists in time by occurring or happening, has temporal parts, and always depends on some (at least one) *material entity*. The dependence here is analogous to that which we find in the relation between a specifically dependent continuant and its independent continuant bearer(s). Examples of BFO: *process* include the life of this organism, that process of meiosis, the course of this disease, that flight of that bird, this process of cell division, this fall of water down this waterfall. My headache (experience of pain in my head) is dependent on me. It cannot also be *your* headache. I can feel your pain, but only in the sense that you and I may experience qualitatively identical pains. But the pain experiences themselves will be numerically distinct; they will be two distinct instances of the same type of pain experience.

The first key feature of processes is that they have temporal parts. Whereas John, considered as a substance, exists along with all of his parts at every instant in which he exists at all, there is no instant of time at which the process we call "John's life" would exist as a whole. Rather, this process unfolds along (and is divided out among) a series of temporal parts, such as for instance John's childhood, his adolescence, his adulthood, his old age; the first year of his life; the seven-thousandth minute of his life, and so forth. These are all temporal parts of John's life, and reflect the fact that John's life can be partitioned into temporal parts in different ways and at different levels of granularity.

Processes such as John's life thus have many other processes as parts. Some are temporal proper parts of John's life (for example, the process that is the sum total of what happens to John during his childhood); some are temporally coextensive with John's life (for example, the process of change in John's temperature from the beginning to the end of John's existence).[1] Each of these processes also has its own existence extended in time, and hence its own temporal parts.

Just as there are relational qualities, so also there are relational processes, which depend on multiple material entities as their relata. For example, John courting Mary, a moving body's crashing into a wall, a game of snooker, the videotaping of an explosion, a war.

Objects, as we have seen, can gain and lose parts while maintaining their identity. In the case of processes, in contrast, the gain and loss of parts is ruled out as a matter of necessity. This is because, if two processes should differ with regard to even the smallest part, then these two processes are nonidentical. John is still John (still numerically the same John) even if he suffers the loss of his arm in an industrial accident; but the process that is John's life and in which he loses his arm is *not the same life as* the process that is John's life (in what we might think of as some alternative possible world) in which he does not lose his arm. John can survive as the same individual (continuant) across many different changes to his parts and qualities, but there is only one life that is his, and this is so independently of whether he can choose how to live this life as it unfolds.

BFO: History

BFO: *histories* are one important subtype of *process*. Each material entity and each site has from the BFO point of view a unique *history*, which is defined as follows:

history = def. the sum of all processes taking place in the *spatiotemporal region* occupied by the material entity or site in question

A history (occurrent) is thus always a history of something (continuant). What is interesting about histories is that they are, in an important sense, complete. For example, the history of John is the sum of all processes that have occurred within John throughout the course of his entire life, at all granularities. Thus the history of an object such as John is more than just the totality of events that might be described in John's biography. It will include also, for example, all the movements of neutrinos within his interior as they pass through, the movements of his blood cells, as well as the movements of his heart and lungs and of all other constituent organs of his body, and so forth.

The relation between a material entity and its history is one-to-one. Histories are thus very special kinds of processes, since not only is it the case that, for any material entity or site, there is exactly one process which is its history; it is also the case that (by definition of BFO: *history*) there is for every history exactly one material entity or site that it is the history of.

BFO: Process Boundary

A BFO: *process boundary* is an occurrent entity that is the instantaneous temporal boundary of a process. Process boundaries are the beginnings and endings of the processes they bound. More precisely: a process boundary is a temporal part that itself has no temporal parts (where the relation of temporal parthood is as defined in chapter 7). It is the limiting or smallest temporal process part. Examples include the forming of a synapse, the onset of REM sleep, the detachment of a finger in an industrial accident, the final separation of two cells at the end of cell-division, the incision at the beginning of a surgery, where all these terms ("forming," "incision," etc.) are understood as referring to instantaneous changes rather than to the results of such changes. (We leave open the question whether process boundaries—like object boundaries—are always fiat in nature, and also whether they manifest something like the granularity dependence that we identified in the realm of continuants.)

BFO: Spatiotemporal Region

A *spatiotemporal region* is an occurrent entity at or in which occurrent entities can be located. A spatiotemporal region is part of spacetime (that is: it is a part of the whole of spacetime) and each spatiotemporal region is defined relative to some frame of reference involving a four-dimensional system of coordinates. Just as BFO, in its continuant representation of entities, views space as a container within which objects and their qualities exist, so the accompanying occurrent representation of processes views spacetime as an analogous container, within which processes unfold and in which spatiotemporal regions can be identified as parts. Examples of such spatiotemporal regions include the region occupied by a human life, the region occupied by the development of a cancer tumor, the region occupied by a process of cellular meiosis, or the region occupied by a war.

BFO's occurrent ontology, in its current version, thus views spacetime, as a whole, existing in its entirety in its four (three plus time) dimensions. Processes, in this spacetime, have a duration, a beginning, and an end. One can think of each process as a temporally extended continuum, a spacetime *worm*, stretched out in and through the single unified container that is the entirety of spacetime. We note that this view of spacetime worms is distinct from popular four-dimensionalist views according to which objects (such as molecules or people or planets) would themselves be extended in time and would have temporal parts. BFO does indeed embrace a four-dimensionalist perspective; but it combines this with a three-dimensionalist perspective for continuants, and does not attempt to reduce the one to the other.

BFO: Temporal Region

A *temporal region* is an occurrent entity that is a part of time (of the whole of time). Temporal regions differ from spatiotemporal regions in that they are extended or serve as boundaries only along the temporal dimension. A temporal region is the result of projecting a spatiotemporal region onto this temporal dimension.

Temporal regions are introduced in BFO to provide a basis for consistent representation of temporal data. Since there is no absolute time, temporal regions—like spatial regions—require for their representation some selected frame of reference. Users of BFO 2.0 are thus encouraged to specify the temporal coordinate system they are using, but this will—in all the applications known to us currently—be either identical to or trivially intertranslatable with the coordinate systems employed by other users (thus with the clock and calendar systems used for keeping track of terrestrial time).

Since temporal regions have temporal parts (are extended in time) in just the way that processes have temporal parts, they belong to BFO's occurrent perspective. Reference to temporal regions is however employed also when referring to BFO's continuant entities, for example, when we use them as a means for indexing relations between continuant entities such as parthood that hold only at certain times.

BFO: Zero-Dimensional Temporal Region

A *zero-dimensional temporal region*—also called a *temporal instant*—is a temporal region that is without extent. For all intents and purposes a zero-dimensional temporal region is a smallest instant of time just as a process boundary is a smallest temporal part of a process. Zero-dimensional temporal regions are the temporal regions that process boundaries are located in. Examples include right now, the moment at which a finger is detached in an industrial accident, the moment at which a child is born, the moment of someone's death, and the turn of the nineteenth century.

BFO: One-Dimensional Temporal Region

A *one-dimensional temporal region*—also called a *temporal interval*—is a temporal region that is extended in time. It has further temporal regions as parts, including its zero-dimensional temporal region boundaries. One-dimensional temporal regions are the temporal regions in which processes occur or unfold. For example, the temporal region that is the first hour of the day, or the nineteenth century, or the temporal region in which John's life is located, or the temporal region occupied by World War II.

An Example of Occurrent Classification

Having outlined the occurrent perspective of BFO, we can now give a simple illustration of BFO's classificatory power by considering how it classifies the entities that will be involved when a woman undergoes an electrocardiogram (EKG/ECG) at a cardiology clinic:

- the EKG test itself is an **instance of** *process,*
- the start and end of the test are **instances of** *process boundary,*
- the specific electrical activity measured by the test is an **instance of** *process,*
- the points in time at which the EKG test starts and ends are **instances of** *temporal instant,*
- the time taken by the test as a whole is an **instance of** *temporal interval*
- any slice of spacetime during the EKG test, for example as represented on the output graph, is an **instance of** *spatiotemporal instant*

Classifying Universals with BFO

As we noted at the beginning of chapter 5, the categories and relations recognized by the BFO continuant and occurrent perspectives can be used to talk about both universals and particulars. An ontology is by our definition a representational artifact whose representational units are intended to represent universals and relationships among universals on the side of reality, but we come to know what universals and relationships exist only by examining particular instances that we observe in reality, for example, in the context of a scientific experiment. And it is not only domain ontologies that are representations of universals (and, by extension, of their particular instances), but also formal ontologies such as BFO. Hence *temporal region*, like other BFO categories, has instances in reality, such as this five-minute interval starting now, and that five-year interval ending last midnight. You yourself are an instance of BFO: *object*. From the BFO perspective, if we are given a universal that current science tells us exists—has instances—on the side of reality, the first question we need to ask is whether these instances are *continuant* or *occurrent* entities. If the universal in question has *continuant* instances, then the next question is whether these are *independent continuant* instances or *dependent continuant* instances, and so on, until the appropriate formal ontological category has been located within one or other of the two BFO hierarchies presented in chapters 5 and 6. It should be noted that the formal-ontological relationships that obtain between different ontological categories will imply also relationships among the instances of these categories. For example, if BFO: *quality* is dependent upon BFO: *material entity*, then every instance of the quality *red* is dependent upon some instance of *material entity* to serve as its bearer.

Exhaustiveness of BFO Categories

BFO is an ontology that is designed to support information-driven scientific research, and itself shares some of the features of an empirical scientific theory. Thus BFO changes (albeit very slowly) in reflection of lessons learned through use, and it will continue to change in the future. Thus it is possible that there are domain-neutral universals (types of entities) in reality that are needed to perform an adequate job of annotating the results of scientific experiments that BFO has thus far failed to incorporate.

Consistent with BFO's principle of fallibilism, we acknowledge that it is possible that future research in ontology and in the natural sciences, as well as continued attempts at specific domain implementations, will reveal the need not only for an expansion of the top-level categories of BFO but also for corrections of its treatment of the universals already recognized. Such corrections have been made already in the development of earlier versions.[2] Clearly, changes in an ontology such as BFO that is used by a large number of independent groups must be managed on the basis of a careful scientific review process involving collaboration between end users and ontology developers and providing documentation of the principled reasons for any proposed changes.

BFO's Perspectivalism

We have now reached the point where the perspectivalism underlying BFO can be more clearly stated. The *continuant* perspective of BFO represents some portion of space and its *continuant* occupants—including qualities of these objects—as they exist at given instants of time. But it does this in such a way that the identity over time of regions of space, and of material entities occupying such regions, and of qualities and other dependent continuants, can be asserted. In this way BFO avoids any reductionist view of continuants as mere sums of object slices or object stages. Time is in a sense external to the continuant perspective, and an assertion to the effect that a given material entity has a given quality at a given time, or that a given material entity is a part of another material entity at a given time, is represented not by referring to the temporal regions involved as extra entities, but rather by using temporal indexing of the pertinent relational verb.

BFO's *occurrent* perspective, by contrast, represents regions of time, and of spacetime, and the processes that occupy them, as if they were being viewed from the perspective of an idealized observer who is assumed to be outside of time. Time is thus internal to the occurrent perspective—the observer discovers that processes can be ordered along the temporal dimension and that they occupy successive temporal regions, the latter being represented explicitly as extra entities. On this view both times and the changes that occur in these times are represented. The occurrent perspective thereby captures the continuous flow of processes each blending into the next, with process parts being distinguishable within larger process wholes on successively finer levels of granularity. The transformation of Jill's hair from blond to brown can be represented as an occurrent process, involving various part processes (changes in color of individual hairs, chemical processes in each hair shaft, processes of cuticle penetration, and so forth).

But it is also possible to represent the passage of time and the occurrence of change from the continuant perspective. This is done by lining up a series of representations of a given portion of reality as it exists at a corresponding series of different times, and then observing the differences and similarities between the objects represented. Such representations can take account also of changes in qualities. Thus we might have a continuant ontology including a representation of Jill (*object*) from last year when she had blond hair (*quality*), and another *continuant* ontology including a representation of Jill as she is this year, when her hair is brown. One could then point to the difference in hair color as a change in quality, but nevertheless identify the object in which this hair color inheres, namely Jill (or Jill's hair), as numerically identical in the two cases.

Using anatomy and physiology as exemplars, we can say that the continuant perspective corresponds to anatomy, the study of the three-dimensional kinds of structures inside the body, while the occurrent perspective corresponds to physiology, the study of the kinds of processes in which these structures participate. And if we can imagine that there is a single representation of physiology for a given organism

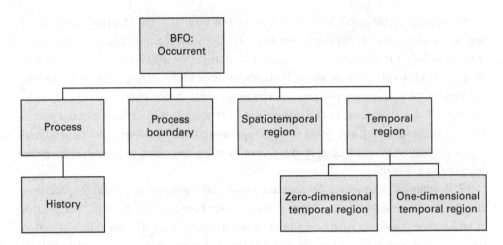

Figure 6.1
The hierarchy of BFO occurrents

extending across the entire set of processes constituting the organism's life, then we must contrast this with the need for a series of distinct anatomical representations as the continuant structures making up the organism change from one life stage to the next—for example, from embryo to fetus to child to adult, and so forth.

Thus the BFO ontology is perspectival along two major dimensions, of continuants and occurrents, respectively. The former correspond to the three-dimensionalist doctrine favored by some contemporary philosophers, the latter to the four-dimensionalist doctrine favored by their opponents. In BFO these two doctrines are combined as alternative, mutually compatible perspectives within a single framework, which incorporates also a corresponding division between two different sorts of granularity, along the continuant and occurrent dimensions, respectively. Just as molecules and cells are at a lower granular level than planets, so the lives of single cells and single organisms are at a lower granular level than entire geological epochs.

BFO's Perspectivalism in Practice

The OBO Foundry[3] is a collaborative experiment in which some dozen ontologies are thus far involved, including the Gene Ontology (GO) at its core. The OBO Foundry is based on the voluntary acceptance of an evolving set of principles of good practice in ontology development by its participants. These include the requirement that ontologies:

Ontological category Granularity	BFO: Continuant				BFO: Occurrent
	Independent		Dependent		
Organ and organism	Organism (NCBI taxonomy or similar)	Anatomical entity (FMA, CARO)	Organ function	Phenotypic quality (PATO)	Biological process (GO)
Cell and cellular component	Cell (CL, FMA)	Cellular component (FMA, GO)	Cellular function	Disease (OGMS, DO, HPO)	
Molecule	Molecule (ChEBI, SO, PRO)		Molecular function (GO)	Phenotype quality (PATO)	Molecular process (MPO)

Figure 6.2
Organization of the OBO Foundry ontologies (with the three branches of the Gene Ontology shaded)

- serve as controlled vocabularies to ensure the accumulation and comparability of scientific research,
- demonstrate usefulness in the annotation and integration of data resources, and
- be semantically interoperable.

The ontologies in the OBO Foundry suite are designed also to bring the benefits of modular development, with collaborating groups of experts taking responsibility for the representations of the entities and relations in their respective domains of expertise.

BFO provides the common upper-level ontology architecture for all the OBO Foundry ontologies, and it thereby also provides the framework within which we can understand the relations between these ontologies as they are developed by separate teams working in tandem. Here GO's three constituent ontologies of Cellular Components (independent continuants), Molecular Functions (dependent continuants), and Biological Processes (occurrents) are mapped within a framework defined in terms of BFO's categories along the horizontal dimension and in terms of levels of granularity along the vertical as in figure 6.2.

Further Reading on Processes and Events

Dretske, Fred. "Can Events Move?" *Mind* 76 (1967): 479–492.

Galton, Anthony. *Qualitative Spatial Change*. Oxford: Oxford University Press, 2001

Grenon, Pierre, and Barry Smith. "SNAP and SPAN: Towards Dynamic Spatial Ontology." *Spatial Cognition and Computation* 4 (1) (2004): 1–10.

Sider, Ted. *Four-Dimensionalism: An Ontology of Persistence and Time*. Oxford: Oxford University Press, 2005.

Simons, Peter. "Continuants and Occurrents." *Proceedings of the Aristotelian Society* 74 (2000): 59–75.

Smith, Barry. "Classifying Processes: An Essay in Applied Ontology." *Ratio* 25 (4) (2012): 463–488.

Zemach, Eddy. "Four Ontologies." *Journal of Philosophy* 23 (1970): 231–247.

7 The Ontology of Relations

In chapter 6, we introduced the basic categories of BFO: *continuant* and BFO: *occurrent*, and their respective subtypes. In this chapter we will introduce the central ontological relations in BFO, and provide examples of how definitions for such relations are to be formulated.

BFO Relations

As has been noted in earlier chapters, providing definitions of the terms representing universals and defined classes alone is normally not sufficient to capture all of the important scientific information about a given domain. The relations that obtain between and among them need to be defined also, and we further need to provide axioms, for example, representing how specific categories are related to each other within the ontology. Definitions and axioms can then be combined together for purposes of reasoning.

Many of our principles of ontology good practice—for example, the principle of single inheritance and of Aristotelian definitions—draw on the central architectural role of the *is_a* relation in ontology construction. Some relations, such as identity and parthood, are primitive; they are so basic to our understanding of reality that it is impossible to conceive of there being anything more basic in terms of which to define them. Here axioms are indispensable if the terms in question are to play more than a trivial role in reasoning about entities in the domain. BFO also includes other relations, such as instantiation, identity, parthood (including both *part_of* and *has_part*), dependence (including both *generic* and *specific dependence*), *located_in*, and a series of further relations pertaining to space and time.[1]

As we discussed in chapter 1, there are three basic kinds of relations that need to be taken into account when designing an ontology and defining the relations that it will represent. These are

• relations holding between one universal and another (the relations represented in the ontology itself);

• relations holding between a particular and a universal—for example, the relationship of *instantiation*, which comes into play where the ontology is applied to some specific portion of reality, for instance in annotating clinical data pertaining to a specific group of patients;

• relations holding between one particular and another—for example, when asserting that Mary's leg is a continuant part of Mary.

Having these three kinds of relations at our disposal allows us to use an ontology in conjunction with information about particulars in the world to reason about those particulars. A paradigm case of this in biomedicine would be a software tool that could allow domain-specific ontologies of biology and medicine to help in guiding decisions as to diagnosis and treatment of specific patients.

It is also important when defining relations to specify what categories of objects form the domain and range of the relation (or in other words what are valid expressions to figure as its left- and right-hand terms, respectively). For example, the relation *instantiates* always holds between a particular and a universal, as in Fido *instantiates* Labrador Retriever. The parthood relation, on the other hand, comes in two forms, the first of which holds between two particulars, the second between two universals. Because these relations behave differently according to whether they obtain between continuants or occurrents, BFO distinguishes further between continuant parthood and occurrent parthood, as we shall explain in more detail.

Relations: Formal Properties and Conventions

First we need to introduce some conventions that will enable us to define the relations between and among universals and particulars:

• the upper-case variables C, D, . . . will be used to represent continuant universals

• the lower-case variables c, d, . . . will be used to represent continuant particulars

• the upper-case variables P, Q, . . . will be used to represent occurrent universals

• the lower-case variables p, q, . . . will be used to represent occurrent particulars

• a relation that holds between two universals will be represented in *italics*, as in: C *is_a* D; P *is_a* Q

• a relation holding between a particular and a universal will be represented in bold-face, as in: c **instance_of** C; p **instance_of** P

Box 7.1
Three Major Families of Relations

---

1. universal-universal
continuant examples

    cancer *is_a* disease   disease *is_a* disposition

    nosocomial infection *is_a* infection   object *is_a* independent continuant

occurrent examples

    meiosis *is_a* cell division   active transport *is_a* membrane transport

    breathing air *is_a* respiration   process *is_a* occurrent

2. particular-universal
continuant examples

    this cell **instance_of** cell   this red (here, of this ball) **instance_of** red

    this myelomeningocele (here, in this girl) **instance_of** myelomeningocele

occurrent examples

    this waltz (being danced here, in Palermo) **instance_of** waltz

    this process of tanning **instance_of** slowing the putrefaction of skin

3. particular-particular
continuant examples

    this atom of hydrogen **continuant_part_of** this water molecule

    this portion of helium gas **continuant_part_of** the sun

    this ion channel **continuant_part_of** this cell membrane

occurrent examples

    this rupturing of ovarian follicle **occurrent_part_of** this process of ovulation here in this fawn

    this process of gamma-glutamylcysteine synthesis **occurrent_part_of** this process of glutathione synthesis

---

• a relation obtaining between two particulars will also be written in bold type, as in: c **continuant_part_of** c at t; or: p **occurrent_part_of** q

The three families of relations and the conventions for representing them are summarized in box 7.1.

Consider the case of Fido and the parthood relation obtaining between Fido (c) and his tail (d). The information communicated by "Fido's tail is a part of the dog Fido" could be represented formally using the conventions just described as follows:

• d **continuant_part_of** c

• d **instance_of** *tail*

• c **instance_of** *dog*

Note that Fido might at some point in time lose his tail, and we will need to address this temporal feature of parthood for continuants. Note, too, that our discussion of Fido here is an example provided only for purposes of illustration. BFO does not assume that *dog* is a universal—since BFO is a domain-neutral ontology that leaves it to biologists to construct domain-specific ontologies in its terms. As already mentioned, an influential school of thought in biology holds that species such as *dog* are more properly to be conceived as evolving dynamic populations of organisms,[2] and a view along these lines could be formulated using the BFO category *object aggregate*. Even then, however, the formulation of such a view would still need to utilize terms designating universals (such as *organism, sexual reproduction, population,* and so forth) to capture what is involved in being a member of a species population, and for these terms the more traditional relation of instantiation between particulars and universals would still be required.

Primitive Instance-Level Relations

We have noted that the categories of any ontology should represent universals in reality. Yet, we will not be able to define what it means for one universal to stand in some relation to another universal—for example, in some parthood relation—without consideration of the underlying relations among instantiations of those universals on the side of particulars. We will show in what follows how universal-universal relations are to be defined in terms of previously accepted primitive relations of the particular-particular and particular-universal sort. A key part of the strategy for understanding universal-universal relations will be to interpret them as being true only if certain other things are true of their instances. Thus we will understand

- phototransduction *occurrent_part_of* visual perception,
- portion of carbon *continuant_part_of* portion of glutathione, and
- phospholipid bilayer *continuant_part_of* mitochondrion

to be true *only if* each particular instance of the universals *phototransduction, portion of carbon,* and *phospholipid bilayer* stands in an instance-level part relation to a corresponding instance of the universals *visual perception, portion of glutathione,* and *mitochondrion,* respectively.

Some of our definitions of relations will involve reference to spatial or temporal regions. Above all, assertions of relations involving continuant particulars will need to include a reference to time. This is so because continuants may change their relations to other entities from one time to the next. We will use the following:

- variables r, r′, . . . to represent three-dimensional spatial regions
- variables t, t′, . . . to represent instants of time

We can now identify the following primitive instance-level relations and their definitions:

- c **instance_of** C **at** t. This is a primitive relation obtaining at a specific time between a continuant instance c and a continuant universal C when the former instantiates the latter at that time. For example: Fido **instance_of** *Labrador Retriever* **at** the present.

- p **instance_of** P. This is a primitive relation obtaining between a process instance and a process universal that it instantiates. (This relation holds independently of time.) For example: John's life **instance_of** *human life*.

- c **continuant_part_of** d **at** t. This is a primitive relation obtaining between two continuant instances and a time at which the one is part of the other. For example: this cell nucleus **continuant_part_of** this cell **at** the present.

- p **occurrent_part_of** q. This is a primitive relation of parthood, holding independently of time, between process instances when one is a subprocess of the other). For example: this tumor's growth **occurrent_part_of** Mary's life.

- r **continuant_part_of** r′. This is a primitive relation of parthood, holding independently of time, between spatial regions (one a subregion of the other). For example: the spatial region occupied by the surface of the Northern hemisphere **continuant_part_of** the spatial region occupied by the whole surface of the earth.

- c **inheres_in** d **at** t. This is a primitive relation obtaining between a specifically dependent continuant and an independent continuant at a particular time. For example: the shape of John's body **inheres_in** John **at** July 26, 2006.

- c **located_in** r **at** t. This is a primitive relation between a continuant instance, a spatial region that it occupies, and a time. For example: John **located_in** the region occupied by the dining room **at** dinnertime.

- r **adjacent_to** r′. This is a primitive relation of proximity between two spatial regions. For example: Northern hemisphere **adjacent_to** Southern hemisphere.

- c **derives_from** d. This is a primitive relation between two distinct material continuants when one succeeds the other across a temporal divide. For example: this blastocyst **derives_from** this ovum.

- p **has_participant** c. This is a primitive relation between a process and a continuant. For example: John's life **has_participant** John.

While these instance-level relations cannot be defined, their meanings can be elucidated informally through the provision of examples, and formally by adding axioms.

For example, having accepted the instance-level relation of continuant parthood (c **continuant_part_of** d **at** t), it is possible to specify its logical properties by explicitly adopting axioms such as

if c **continuant_part_of** d **at** t, then c and d **exist at** t, and

if c **instance_of** *continuant* then there is no d such that c **occurrent_part_of** d **at** t,

and so on.

Universal-Universal Relations in BFO

In the previous section, we considered some of the primitive instance-level relations that scientists—with or without realizing it—draw on in their work. In order now to define what it means for one universal to stand in relation to another we need to consider the particular instantiations of those universals. This should not be seen as standing in conflict with our view that ontologies are representational artifacts whose representations are intended to designate some combination of *universals* and the relations between them. For in defining the relations between universals, the reference to particulars is entirely general—we will be referring in effect to *all* particulars instantiating given universals. This corresponds to the way in which science as a whole concerns itself with *generalities* when formulating its types, laws, and relations. It tests its hypotheses concerning such generalities by examining particulars in experiments; but references to specific particulars do not play a role when representing scientific laws.

In 2005 Smith, together with a group of influential researchers in the biomedical ontology field,[3] compiled a list of ten basic universal-universal relations under the categories of

- foundational relations
- spatial relations
- temporal relations
- participation relations

proposing them as a common basis for the further development of biomedical ontologies in separate disciplinary communities. These relations have to a large degree been reused in the OBO Foundry and in a number of related ontologies (see box 7.2), while at the same time the list has been expanded with further relations also recommended for common use.[4] Work is also ongoing in the context of the formulation of the OWL version of BFO 2.0 to create a complete system of formal definitions of all the relations between BFO categories.

Box 7.2
Core Relations in BFO

Foundational Relations

1. *is_a* (is a subtype of)
   - *portion of deoxyribonucleic acid (DNA) is_a portion of nucleic acid*
   - *photosynthesis is_a physiological process*
2. *continuant_part_of*
   - *cell nucleus continuant_part_of cell*
   - *heart continuant_part_of cardiovascular system*
3. *occurrent_part_of*
   - *neurotransmitter release part_of synaptic transmission*
   - *gastrulation part_of animal development*

Spatial Relations

4. *located_in*
   - *intron located_in gene*
   - *chlorophyll located_in thylakoid*
5. *adjacent_to*
   - *Golgi apparatus adjacent_to endoplasmic reticulum*
   - *periplasm adjacent_to plasma membrane*

Temporal Relations

6. *derives_from*
   - *mammal derives_from gamete*
   - *triple oxygen molecule derives_from oxygen molecule*
7. *preceded_by*
   - *translation preceded_by transcription*
   - *digestion preceded_by ingestion*

Participation Relations

8. *has_participant*
   - *death has_participant organism*
   - *breathing has_participant thorax*

In the rest of this chapter, we will examine these relations as well as provide definitions (where possible) and examples of each. We shall conclude by providing some examples of axioms illustrating how these relations are treated formally within the larger BFO framework.

Foundational Relation: *is_a*

The foundational relation *is_a* has been discussed at length already. Examples include

- *myelin is_a lipoprotein*
- *beer is_a alcoholic beverage*
- *eukaryotic cell is_a cell*
- *site is_a independent continuant*

for *continuant* entities, and

- *gonad development is_a organogenesis*
- *binge drinking is_a drinking*
- *intracellular signaling cascade is_a signal transduction*

for *occurrent* entities.

We define the *is_a* relations in terms of the primitive relation **instance_of** introduced previously. For occurrents this reads as follows:

A *is_a* B = def. A and B are universals and for all x (if x **instance_of** A then x **instance_of** B)

The corresponding definition for continuants is created by using the temporally qualified relation **instance_of at t**, as in the definition of *continuant_part_of* on p. 139.

Thus *diploid cell is_a cell* means: for any particular continuant c, c **instance_of** *diploid cell* **at** t implies c **instance_of** *cell* **at** t. *Lung cancer development is_a cancer development* means: for any particular occurrent p, p **instance_of** *lung cancer development* implies p **instance_of** *cancer development*.

Foundational Relations: *continuant_part_of* and *occurrent_part_of*

BFO distinguishes two foundational parthood relations namely *continuant_part_of* and *occurrent_part_of*. Examples include

- *axon continuant_part_of neuron*
- *cell nucleus continuant_part_of cell*
- *neuronal death occurrent_part_of dementia*
- *bird song occurrent_part_of avian mating behavior*

These relations can be defined in terms of the instance-level parthood relations as follows:

C *continuant_part_of* D = def. for every particular continuant c and every time t, if c **instance_of** C **at** t, then there is some d such that d **instance_of** D **at** t and c is a con**tinuant_part_of** d **at** t

So, for example, to say that *cell nucleus continuant_part_of cell* is to say that for every particular cell nucleus and every time in which that cell nucleus exists, there is some instance of *cell* which this cell nucleus is an instance-level **continuant_part_of** at that time. Notice that this does not require that every instance of cell have some instance of nucleus as part.

For occurrent universals we have

P *occurrent_part_of* Q = def. for every particular occurrent p, if p **instance_of** P, then there is some particular occurrent q such that q **instance_of** Q and p **occurrent_part_of** q

Thus, for example, *human neurulation occurrent_part_of human fetal development* just in case every particular process of human neurulation is an instance-level **occurrent_part_of** some particular process of human fetal development.

Spatial and Temporal Relations

One important influence on the development of BFO was the Region Connection Calculus (RCC), a simple framework to support qualitative reasoning about spatial relations[5] and currently integrated into the GeoSPARQL standard for representation and querying of geospatial linked data for the Semantic Web.[6] The Allen Interval Algebra is an analogous and similarly influential framework for reasoning about temporal relations.[7]

In what follows we provide sample definitions of spatial and temporal relations in BFO, which are designed to serve as a basis for defining all the relations defined in RCC and in the Allen calculus—and further analogous spatiotemporal relations according to need. A particularly ambitious set of such relations for spatial adjacency and connectedness is defined in the BFO-conformant Foundational Model of Anatomy (FMA) Ontology.[8] We provide here a description of the treatment of location and adjacency in conformance with BFO, which is formulated in terms of the relations between the spatial regions that independent continuants occupy. The strategy employed here can then be generalized to spatial relations of other sorts.

Both *located_in* and *adjacent_to* connect one spatially extended entity to another in terms of the relations between the spatial regions (r, r', . . .) that they occupy. Examples of *located_in* include

- *ribosome located_in cytoplasm*
- *Golgi body located_in cell*

Defining the universals-level relation *located_in* comes in two steps. First, we introduce the instance-level primitive relation c **located_in** r **at** t (obtaining between a continuant instance, a spatial region, and a time). Second, we define an instance-level relation (obtaining between two continuant particulars and a time) c **located_in** d **at** t, and then use this relation to define the *located_in* relation at the level of universals.

c **located_in** d **at** t = def. there are two spatial regions r and r′ such that the particular continuant c is **located_in** r **at** t and the particular continuant d is **located_in** r′ **at** t, and the region r is a **continuant_part_of** the region r′

For example, John's kidney is **located_in** John's torso at present because the region in which John's kidney is **located** is a **continuant_part_of** the region in which his torso is located.

   Given this definition, we can now define

C *located_in* D = def. for every particular continuant c and every time t, if c **instance_of** C **at** t, then there is some d such that d **instance_of** D **at** t and c **located_in** d **at** t

Thus, *kidney located_in torso* means that, for every (instance of) kidney and for every time t at which that kidney exists, there is some instance of torso at that time such that the kidney is **located_in** the torso at that time.

   As we have already seen in our discussion, for example, of bacteria, while all continuant parts of spatially extended entities are located in those entities not all continuant entities located in the interiors of spatially extended entities are parts thereof. Note that (for example, because of kidney transplants) great care must be taken when incorporating assertions such as *kidney located_in torso* into an ontology. In the FMA, the problem is solved by treating assertions of this sort as holding, not of the actual human anatomy instantiated by you or me, but rather of what is called *canonical* human anatomy, defined as the arrangement and structure of body parts that is generated by the coordinated expression of the structural genes of the human organism.[9]

Spatial Relation: *adjacent_to*
The relation *adjacent_to* is a relation of proximity between disjoint continuants. Examples include the following:

- *nuclear membrane adjacent_to cytoplasm*
- *seminal vesicle adjacent_to urinary bladder*
- *ovary adjacent_to parietal pelvic peritoneum*

This relation can now be defined in similar pattern:

C *adjacent_to* D = def. for every particular continuant c and for every time t, if c **instance_of** C at t, then there is some d such that d **instance_of** D **at** t and c **adjacent_to** d **at** t

Thus, *liver adjacent_to falciform ligament* means that, for every instance of *liver* and every time t, if the liver exists at t, then there exists an instance of *falciform ligament* at t, and the liver is **adjacent_to** the falciform ligament **at** t. **adjacent_to** for material entities can itself be defined in terms of the relation of adjacency between the regions they occupy as defined in RCC.

Temporal Relation: *derives_from*

The temporal relation *derives_from* is used to assert that each instance of one continuant universal is derived from some instance of a second universal. Different *derives_from* relations have been proposed by biologists working in different disciplines. The relation we consider here interprets the relation to hold in those cases where a biologically significant portion of the matter contained in the earlier instance is inherited by the later instance. Examples include the following:

- *plasma cell derives_from B lymphocyte*
- *portion of tyrosine derives_from portion of henylalanine*

The underlying instance-level **derives_from** relation in cases of this sort can be understood as meaning that c is such that in the first moment of its existence it occupies a spatial region that substantially overlaps the spatial region occupied by d in the last moment of its existence. We can then define:

C *derives_from* D = def. for every particular continuant c and every time t, if c **instance_of** C at t, then there is some d and some earlier time t′ such that d **instance_of** D at t′, and c **derives_from** d

We can think of the relation so defined as *immediate* derivation, and define from there different sorts of *mediated* derivation (so that we could infer, for example, from C *derives_from* D and D *derives_from* E that C *mediately_derives_from* E).

Temporal Relation: *preceded_by*

The temporal relation *preceded_by* is used in assertions such as

- *translation preceded_by transcription*
- *aging preceded_by development*
- *neurulation preceded_by gastrulation*

The underlying instance-level **preceded_by** relation is to be understood in the obvious way (and in conformity with the Allen calculus) as meaning that the temporal region occupied by the process p is later than the temporal region occupied by the process q.

Preceded_by as a relation between occurrent universals can then be defined as follows:

P *preceded_by* Q = def. for every particular occurrent p, if p **instance_of** P, then there is some q, such that q **instance_of** Q and p **preceded_by** q

Participation Relation: *has_participant*

The relation *has_participant* holds between a process and a continuant entity when the latter participates in or is involved in the former. Examples include the following:

• *cell transport has_participant cell*
• *translation has_participant portion of amino acid*
• *cell division has_participant chromosome*

Thus, every instance of *cell transport* (occurrent) has some instance of *cell* (continuant) as participant, every instance of *translation* (occurrent) has some instance of *portion of amino acid* (continuant) as participant, and so on. We can accordingly define the following:

P *has_participant* C = def. for every particular occurrent p, if p **instance_of** P, then there is some c and some time t such that c **instance_of** C at t and p **has_participant** c at t

Some Further Top-Level Relations

We have now described the primitive-level instance relations and universal-universal relations that are recognized by BFO, providing both definitions (where possible) and examples. We now consider some additional relations designed for use in specific domains and proposed for inclusion in the Relation Ontology (RO).[10]

proper_continuant_part_of and ***proper_occurrent_part_of***

To speak of a "proper part of" something at the instance level is to speak of something that is a part of but not identical with something else. With respect to continuant universals, an example is this: human uterus *proper_continuant_part_of* human. We define as follows:

C *proper_continuant part_of* D = def. for every particular continuant c and every time t, if c **instance_of** C at t, then there is some d such that d **instance_of** D at t, and c **proper_continuant_part_of** d at t

With respect to occurrents:

P *proper_occurrent_part_of* Q = def. for every particular occurrent p, if p **instance_of** P, then there is some particular q such that q **instance_of** Q, p **proper_occurrent_part_of** q

Examples are

- *swallowing proper_occurrent_part_of eating*
- *anaphase proper_occurrent_part_of mitosis*

Has_continuant_part and *integral_continuant_part*; *has_occurrent_part* and *integral_occurrent_part*

To assert of two universals that the first has the second as part is to assert that the former has instances that are wholes, and that each such whole has an instance of the latter as part. Thus

C *has_continuant_part* D = def. for all particular continuants c, and all times t, if c **instance_of** C **at** t then there is some D, such that d **instance_of** D **at** t, and d **continuant_part_of** c **at** t

P *has_occurrent_part* Q = def. for all particular occurrents p, if p **instance_of** P then there is some q, such that q **instance_of** Q, and q **occurent_part_of** p

As we shall see, the relation between *has_part* and *part_of* is logically not quite trivial. One can speak of the relation of "integral parthood" holding between two universals A and B when A *part_of* B and B *has_part* A. Thus for the continuant case:

C *integral_continuant_part_of* D = def. C *continuant_part_of* D and D *has_continuant_part* C

and similarly for occurrents

Examples are

- *brain integral_continuant_part_of mammal*
- *systole integral_occurrent_part_of cardiac cycle*

Relations and Definitions of Categories

It is important to note that well-defined formal relations can be used to help more precisely define the nature of the universals that they relate. For example, a universal can be asserted to be an entity that is not an *instance_of* anything, while a particular can be asserted to be an entity that is an **instance_of** other entities, but does not itself have any entities standing in the **instance_of** relation to it. Axioms of these sorts are provided in a section to follow. Similarly, an independent continuant can be asserted to be an entity to which other entities stand in the **inheres_in** relation, but which does not itself **inhere_in** any other entity. Such relational assertions can be used in definitions and thereby help the ontology to support formal reasoning.

The All-Some Rule

Apart from *is_a* all the relations defined as obtaining between universals adhere to what we call the all-some rule. If a universal A bears such a relation to a universal B, then **all** relevant instances of A must bear the relevant instance-level relation to **some** instance of B. This point can be captured simply by saying that relations obtaining among universals should not admit exceptions.

Consider the definition of *continuant_part_of*:

C *continuant_part_of* D = def. for every particular continuant c and every time t, if c **instance_of** C **at** t, then there is some d such that d **instance_of** C **at** t and c is a **continuant_part_of** d **at** t

What this says is that for one universal to be *continuant_part_of* another, *all* instances of the one must at the pertinent times be **continuant_part_of** *some* instance of the other. Thus to say that *human heart continuant_part_of human circulatory system* is to say that for every particular human heart, at every time at which the heart exists there is some instance of human circulatory system which that human heart is part of at that time. Note that it is not implied that it is the same instance of human circulatory system that is involved from one time to the next.

An analogous feature holds of the *is_a* relation in that here, too, we are interested in what holds universally. *Prokaryotic cell is_a cell* is from this point of view in perfect order as it stands. However, *cancer is_a terminal disease* fails to respect the test of universality, since not all instances of *cancer* are instances of *terminal disease*.

Inversion and Reciprocity

As will be clear from our discussion of "has part," it is possible to ask in regard to any relation defined in an ontology whether there is another relation that goes, as it were, in the other direction—generally referred to as the *inverse* relation. The inverse of a relation R is defined as the relation that obtains between each pair of relata of R when taken in reverse order.

So, if C *is_a* D, the relation between D and C that goes in the opposite direction is the relation *has_subuniversal*, defined by

C *has_subuniversal* D = def. D *is_a* C,

as in

cell has_subuniversal prokaryotic cell

(again with suitable modifications when account is taken of defined classes as relata of the *is_a* relation).

However, for most of the relations that we have defined, it is not possible to define an inverse relation in this direct fashion. This makes it necessary to define what have been called *reciprocal* relations.

To see the problem consider the assertion

human testis continuant_part_of human,

which passes the all-some test. When we attempt to formulate the assertion in the opposite direction, however, the result

human has_continuant_part human testis

fails by the all-some test, because not all humans have testes. *Has_continuant_part* is not the inverse but rather what we call the reciprocal of *continuant_part_of.* Something similar holds for the *adjacent_to* relation, for while we have for adult humans

uterine artery adjacent_to urinary bladder

we do not have

urinary bladder adjacent_to uterine artery

since male humans do not have a uterine artery.

The relations *is_a* and *has_subuniversal* allow us to reason both upward and downward through the taxonomical backbone of an ontology, following the principle of inheritance. If A *has_subuniversal* B then everything that holds of every instance of A will hold also of every instance of B. Inferences of this sort are possible also by using other relations in an ontology, but as can be seen from the examples of parthood and adjacency, they must be used with care.

Some Examples of Axioms

Our purpose in this chapter has been to give only an outline of what a theory of relations for purposes of ontology construction within the BFO framework looks like. It is not to develop a full axiomatic theory. However, we here provide for purposes of illustration some examples of axioms for BFO, formulated in English and in first-order logic.[11]

Entity

All *entities* exist at some *temporal region*.

∀x (Entity(x) → ∃t (TemporalRegion(t) ∧ exists_at(x, t)))

Material Entity

Every *material entity* has a *history*.

∀x (MaterialEntity(x) → ∃y has_history(x, y))

Every *entity* that has a *material entity* as continuant part is a *material entity*.

∀y∀x∀t ((continuant_part_of(x, y, t) ∧ MaterialEntity(x)) → MaterialEntity(y))

Every *material entity* exists at some *temporal interval*.

∀x (MaterialEntity(x) → ∃t (1DTemporalRegion(t) ∧ exists_at (x, t)))

Occurrent

Every *occurrent* occupies some *spatiotemporal region*.

∀x (Occurrent(x) → ∃y (SpatioTemporalRegion(y) ∧ occupies_spatiotemporal_region(x, y)))

SpatioTemporalRegion

Every *spatiotemporal region* occupies *some temporal region*.

∀x (SpatioTemporalRegion(x) → ∃t (TemporalRegion(t) ∧ occupies_temporal_region(x, t)))

Every *spatiotemporal region* occupies itself.

∀x (SpatioTemporalRegion(x) → occupies_spatiotemporal_region(x, x))

TemporalRegion

All *temporal regions* are either *zero-* or *one-dimensional* (i.e., either instants or intervals).

∀x (TemporalRegion(x) ↔ (1DTemporaRegion(x) ∧ 0DTemporalRegion(x)))

Every *temporal region* occupies itself.

∀x (TemporalRegion(x) → occupies_temporal_region(x, x))

Universal

Something is a universal if and only if it is instantiated by something.

$\forall X$ (Universal(X) $\leftrightarrow$ $\exists y$ inst(X, y))

Continuant_part_of

The part of relation for *continuants* is anti-symmetric.

$\forall x \forall y \forall t$ ((continuant_part_of(x, y, t) $\wedge$ continuant_part_of (y, x, t)) $\rightarrow$ x = y)

The part of relation for *continuants* is transitive.

$\forall x \forall y \forall z \forall t$ ((continuant_part_of(x, y, t) $\wedge$ continuant_part_of (y, z, t)) $\rightarrow$ continuant_part_of (x, z, t))

The part of relation for *continuants* is reflexive.

$\forall x \forall t$ ((Continuant(x) $\wedge$ exists_at(x, t)) $\rightarrow$ continuant_part_of (x, x, t))

Weak supplementation: If x is a proper (continuant) part of y, then there is some (continuant) part of y that does not overlap x.

$\forall x \forall y \forall t$ (proper_continuant_part_of(x, y, t) $\rightarrow$ $\exists z$ (continuant_part_of(z, y, t) $\wedge$ ¬continuant_overlap(z, x, t)))

Unique product: If one *continuant* overlaps another *continuant* at some time, then there is a unique mereological product (intersection) of those *continuants* at that time.

$\forall x \forall y \forall t$ (continuant_overlap(x, y, t) $\rightarrow$ $\exists z$ (continuant_mereological_product(z, x, y, t) $\wedge$ $\forall w$ (continuant_mereological_product(w, x, y, t) $\rightarrow$ w=z)))

If some *continuant* is part of a *continuant* at some time, then both continuants exist at that time.

$\forall x \forall y \forall t$ (continuant_part_of(x, y, t) $\rightarrow$ (exists_at(x, t) $\wedge$ exists_at(y, t)))

Occupies_spatial_region

Something can only occupy one *spatial region* at a time.

$\forall x \forall r_1 \forall r_2 \forall t$ ((occupies_spatial_region(x, r_1, t) $\wedge$ occupies_spatial_region(x, r_2, t) $\rightarrow$ r_1=r_2)

All entities that occupy a *spatial region* at a time exist at that time.

$\forall x \forall r \forall t$ (occupies_spatial_region(x, r, t) $\rightarrow$ exists_at(x, t))

Box 7.3
Properties of Instance-Level Relations in BFO

Relation	Transitive	Symmetric	Reflexive	Antisymmetric
is_a	+	-	+	+
part_of	+	-	+	+
located_in	+	-	+	-
adjacent_to	-	-	-	-
derives_ from	-	-	-	-
preceded_by	+	-	-	-
has_participant	-	-	-	-

Occupies_spatiotemporal_region

Something can occupy only one *spatiotemporal region*.

$\forall x \forall r_1 \forall r_2$ ((occupies_spatiotemporal_region(x, r_1) $\land$ occupies_spatiotemporal_region(x, r_2)) $\rightarrow r_1 = r_2$)

Reflexivity, Symmetry, and Transitivity

We conclude by summarizing a number of well-understood properties of relations that should be taken into account when defining further relations within the BFO framework. (See Box 7.3.) Here A, B, . . . range over all entities, whether universals, defined classes or particulars.

- To say that a relation R is *reflexive* is to say that anything A that bears the relation R to something else, B, also bears that relation to itself. The relation "is as tall as" is reflexive, because when John is as tall as Jill, he also stands in this same relation to himself: John is as tall as John.

- To say that a relation R is *symmetric* is to say that if A stands in R to B then B also stands in R to A. The instance-level relation **adjacent_to** is symmetric, because if John is next to Mary, then Mary is also next to John. On the level of universals, however, *adjacent_to* is not symmetric.

- To say that a relation R is *transitive* is to say that if a thing A bears R to B, and if B bears R to C, then A also bears R to C. A simple example of a transitive relation is "is taller than." If John is taller than Mary, and Mary is taller than Steve, then John is taller than Steve.

- To say that a relation R is *antisymmetric* is to say that if A bears R to B and B bears R to A, then A and B are identical.

Further Reading on Relations

Bennett, Brandon. V. Chaudhri, and N. Dinesh. "A Vocabulary of Topological and Containment Relations for a Practical Biological Ontology." In *Spatial Information Theory: Proceedings of COSIT 2013*, Lecture Notes in Computer Science, vol. 8116, ed. J. Stell, T. Tenbrink, and Z. Wood, 418–437. Scarborough, UK: Springer, 2014.

Bittner, Thomas, and Maureen Donnelly. "Logical Properties of Foundational Relations in Bio-ontologies." *Artificial Intelligence in Medicine* 39 (2007): 197–216.

Donnelly, Maureen, Thomas Bittner, and Cornelius Rosse. "A Formal Theory for Spatial Representation and Reasoning in Biomedical Ontologies." *Artificial Intelligence in Medicine* 36 (2006): 1–27.

Smith, Barry, Werner Ceusters, Bert Klagges, Jacob Köhler, Anand Kumar, Jane Lomax, Chris Mungall, Fabian Neuhaus, Alan L. Rector, and Cornelius Rosse. "Relations in Biomedical Ontologies." *Genome Biology* 6 (5) (2005). doi:10.1186/gb-2005-6-5-r46. Accessed September 25, 2014.

8 Basic Formal Ontology at Work

All good ontology building, in our view, starts with thinking. The author of an ontology should first assemble a collection of the major terms he will need to use, and use careful thinking to ensure that he understands the meanings of these terms within the context of the associated science, and then further careful thinking to ensure that he can define these terms using words that other human beings will understand and in ways conformant to the BFO ontology and to the other principles set forth in the foregoing.

At some point, however, the ontology builder will need to embark on the process of creating the ontology as a computer artifact, a piece of software that can be used and reasoned with. In the future we believe that a range of different sorts of approaches to the building of such computer artifacts will be available, including an approach based on first-order logic (FOL) along the lines illustrated in the set of sample axioms provided in the previous chapter. FOL, in our view, provides the sort of expressivity that is needed to create formal definitions that will be conformant with the content of biological and other scientific disciplines. Currently, however, the most widely used standard approaches are focused on formal languages that have an expressivity weaker than FOL, primarily the Web Ontology Language (OWL). In this chapter we provide a brief description of the Protégé ontology-building tool and of OWL, which is the primary logico-linguistic framework used in what are called Semantic Web technologies, practitioners of which are a target audience for Protégé. We then provide examples of domain ontologies that utilize BFO as an upper-level ontology in order to give the reader an opportunity to see how the principles and recommendations from the previous chapters are used in practice.

The Protégé Ontology Editor and BFO

In this book, our focus has been on giving researchers an introduction to theoretical principles and on the strategies for creating good ontology content—rather than on the computational tools and details of ontology implementations. That said, we will describe briefly one major public-domain ontology building and editing tool known as Protégé.[1] We will talk also about OWL,[2] which is the primary logico-linguistic framework used in what is called the "Semantic Web" that is supported by Protégé.[3]

Protégé is described at http://protege.stanford.edu/ as a "free, open-source ontology editor and framework for building intelligent systems." The software is freely available for download on any computer, and with the help of tutorials provided at the Protégé website it enables users to start composing domain ontologies. These ontologies are intended, potentially with the help of BFO, to be interoperable with other domain ontologies built in a similar way.[4]

The latest OWL version of BFO can always be found at the following link: http://purl.obolibrary.org/obo/bfo.owl. Once users have downloaded and installed the Protégé software, they can simply import this latest version into a new Protégé file and start constructing an ontology with terms ("classes") relevant to their respective domain by defining these as subclasses of terms in BFO. At various points in this chapter, we will feature screenshots of ontologies that have been produced using this method.

The Web Ontology Language (OWL)

Protégé is a tool that has ontologies constructed using OWL as one of its outputs. OWL evolved from two sources that were merged in the early 2000s: the U.S. Defense Advanced Research Projects Agency (DARPA) *Agent Markup Language* (DAML) and the European Union's *Ontology Inference Layer* (OIL).[5] By 2002 the combined resource (then called "DAML+OIL") had been recommended for use by the World Wide Web Consortium (W3C), which is one of the principal Web standards organizations.[6] The committee approving DAML+OIL was comprised of now-well-known working ontologists such as Ian Horrocks, Peter Patel-Schneider, Jim Hendler, Deb McGuinness, and the person credited with inventing the World Wide Web itself, Tim Berners-Lee. In February 2004 OWL received the status of a recommendation by the W3C with the following abstract, still accurate today: "The OWL Web Ontology Language is designed for use by applications that need to process the content of information instead of just presenting information to humans. OWL facilitates greater machine interpretability of Web content than that supported by XML, RDF, and RDF Schema (RDFS) by providing additional vocabulary along with a formal semantics."[7]

In the next several sections, we will parse the main ideas in this abstract.

Hypertext Markup Language (HTML) and Extensible Markup Language (XML)

HTML was developed by Berners-Lee in the late 1980s. It was designed to allow Web developers to display information in a way that is accessible to humans for viewing via Web browsers.[8]

HTML's primary limitation is that it is oriented toward representations of documents, and so it does not provide the capability to describe information in ways that facilitate the use of software programs to find, interpret, validate, or combine it. Thus, in 1998 the W3C recommended for use the Extensible Markup Language (XML) that allows information to be more accurately described utilizing tags that are not only understandable by humans but also readable and interpretable by machines. In the words of the members of W3C's XML Working Group: "XML is the universal format for structured documents and data on the Web. It allows you to define your own mark-up formats."[9]

Resource Description Framework (RDF)

Unfortunately, while XML is useful when one wants to query documents, it has a limited capability to convey in a standard form the meanings of the statements contained therein, for example, as concerns the relationships between and among entities, and this prompted researchers to develop on its basis the Resource Description Framework (RDF). RDF became a W3C recommendation in February 1999.[10] It is a language created to allow the representation of relationships between entities by means of a simple subject-predicate-object format known as a *triple*. RDF thus to a degree mimics the structure of human language. It also allows for a computational representation of entities and relationships that allows reasoning that is more powerful than can be achieved using XML alone.

The *resource* part of RDF is inspired by the ability provided by the Internet to use Uniform Resource Identifiers (URIs), which is to say standard web addresses, to unambiguously identify web pages on the Internet. In RDF the use of "resource" refers to the fact that all of the subjects, predicates, and objects making up each triple are seen as referring to entities in reality, all of which are called "resources." XML does not represent reality in this way at all, though it does provide specifications for representing dates, numbers, and symbols.

The triples form an RDF database—called a *triplestore*—which can be populated with detailed information about some domain. The subjects, predicates, and objects are tagged with URIs (or in certain cases by what are called "blank nodes"), and the information can be placed on the Web so that anyone in the world can query the database.

RDF Schema (RDFS)

Immediately after RDF began to be used in the early 1990s, RDF Schema (RDFS) was developed as a set of mechanisms for describing groups of related resources (understood, again, in terms of URIs) and of the relationships between them.

Where RDF allows assertions about instances and literals such as integers or alphanumeric strings, RDFS allows also assertions about types: *schematic* assertions. Thus, for example, it allows the representation of the domains and ranges of relations, which in the Semantic Web world are called "properties." For example, it allows the property *authored_by* to be characterized as having a domain *document* and a range *person*. Information of this sort can then be saved in a triplestore and queried with SPARQL (see next section).[11] RDFS introduces a number of predicates including:

- rdfs:label, refers to a string of text describing a resource
- rdfs:comment, points to a human-readable comment about the resource, also often used when providing a definition
- rdfs:seeAlso, which links to other relevant resources
- rdfs:subClassOf, used to state that all the instances of one class are instances of another
- rdfs:domain, used to state that any resource that has a given property is an instance of one or more classes
- rdfs:range, used to specify the set of values that a property can accept

By adding such elements, RDFS was able to augment the expressivity of RDF in order to enable more adequate representation of given domains of interest.

However, almost immediately after RDFS was developed, researchers began to see that it was still too weak to talk about types and about instances belonging to a type, as well as about properties of relations. Further, given that many of these same researchers had been working in the field of artificial intelligence, they wanted a language that could enable machine reasoning.[12] But RDFS is not able, for example, to express the assertion that some relation is transitive in a way that would allow users of RDFS to exploit transitivity in their reasoning.

RDFS is similarly unable to account for the symmetric, reflexive, and other properties of relations that we discussed in chapter 7. It is impossible to say in RDFS, for example, that a relation is symmetrical or that one relation is the inverse of another.

Further, although RDFS added the ability to specify the domain and range of properties, there is no ability to constrain or localize the domain or the range. For example, it is not possible to designate that the range of *has_offspring* is *person* when applied to persons, but *dog* when applied to dogs, *cat* when applied to cats, and so forth. Also, there is no way to express existential or cardinality constraints in RDFS. One cannot say, for example, that all instances of *person* have a *mother* that is also a *person*, or that *persons* have *exactly two biological parents*.[13] These and other problems with RDFS prompted researchers to develop OWL.

Simple Protocol and RDF Query Language (SPARQL)

Just as tuples are queried with Structured (or Standard) Query Language (SQL) in a standard relational database, so triples are queried in a triplestore with what is known as the Simple Protocol and RDF Query Language (SPARQL). SPARQL 1.1 became a W3C recommendation in March 2013.[14]

A simple SPARQL query might look like this:

```
SELECT DISTINCT ?predicate
WHERE { ?subject ?predicate ?object }
ORDER BY ?predicate
```

which asks for all of the predicates in a triplestore.

Basic Features of OWL

In the preceding chapters we have distinguished universals (such as *person* and *role*), whose instances are entities in their own right, from defined classes, which we can think of as devices to capture certain ways of speaking about reality (as, when we speak of lawyers or students, we are in fact speaking merely of *persons* with special *roles*). In the worldview of RDF and OWL, now, universals and defined classes are in effect run together under the single heading of "classes"; and (binary) relations are referred to as "properties." (OWL does share with BFO the use of "instance"; it uses this term to refer to the individuals that are the members of classes as it conceives them.) OWL is in this respect close to a traditional set-theoretic view of reality; however, because it allows for there to be two classes with the same members, OWL classes are in fact intensional.

A key feature of OWL is its basis in Description Logics (DLs), a family of logics that are expressively weaker than standard FOL, but enjoy certain computational properties advantageous for purposes such as ontology-based reasoning and data validation. DLs emerged out of the artificial intelligence community in the mid-1980s and have a formal semantics typically expressed using model theory of the sort familiar from the world of FOL. In contradistinction to XML, RDF, and RDFS, OWL allows for the expression of

a. universal ($\forall$) quantification, through the owl:allValuesFrom restriction;

b. existential ($\exists$) quantification, through the owl:someValuesFrom or owl:hasValue restriction;.

c. cardinality through owl:cardinality, owl:minCardinality, and owl:maxCardinality;

d. the Boolean *and*, *or*, and *not*, which in OWL are called owl:intersectionOf, owl:unionOf, and owl:complementOf, respectively;

e. assertions of equivalence, through owl:equivalentClass and owl:equivalentProperty (thus it is possible to assert for example that two classes X and Y are equivalent, by which is meant that they have the same members [apply to the same instances]); and

f. properties declared to hold of relations, including inverse (owl:inverseOf), functional (owl:FunctionalProperty), inverse functional (owl:InverseFunctionalPrope rty), transitive (owl:TransitiveProperty), symmetric (owl:SymmetricProperty), asymmetric (owl:AsymmetricProperty), reflexive (owl:ReflexiveProperty), and irreflexive (owl:IrreflexiveProperty).

OWL recognizes two types of properties: *object properties* and *data properties*. In conjunction with the designation of a domain and range, an object property (owl:ObjectProperty) specifies a relationship between two individuals (instances, members)—as in, "wing *part_of* airplane" or "wrist *adjacent_to* hand," where "wing," "airplane," "wrist," and "hand" refer to specific instances of the corresponding classes, not to the classes (or universals) themselves. A data property (owl:DatatypeProperty) specifies a relationship between an individual and a literal (integer, double, float, string, boolean, etc.); for example, "the number of participants in the meeting was 6."

To check the consistency of any given set of facts or axioms (including definitions as this term is used in the preceding), and thus in particular to check the consistency of an ontology in OWL, we must use *reasoners* which can determine for each class in the ontology whether it is possible for that class to have instances.[15] A number of reasoners can be used in conjunction with Protégé, which are discussed in the appendix.

OWL vs. Standard Relational Databases

There are three further points to be made about ontologies built in OWL that can best be seen by contrasting the use of such ontologies with the use of standard relational databases.

First, in the latter each instance/individual must have a unique identifier; for example, the planet Venus must be referenced using one name only, such as "Venus," or your grandmother must be referenced using only one identifier, for example, her Social Security number. This is because it is rows in a database that are typically used to represent instances and there is a rule (the primary key uniqueness constraint) that prevents two rows being used to represent the same instance. In an OWL ontology, however, an instance can have more than one name; Venus can be referenced as either "The Morning Star" or "The Evening Star," or both, and one's grandmother can be referenced as "Grandma," "Nana," "Florence Smith," or all three. SameAs axioms are then included (or SameAs relations are inferred) to assert identity.

Second, the closed-world assumption is standardly presupposed by those working with relational databases. The means that *what is not known to be true in the database* is by default considered false, because knowledge of the world represented in the database is assumed to be complete. (This assumption is possible in virtue of the fact that the world of the database is defined exhaustively by the database itself.) Ontologies, by contrast, utilize the open-world assumption (OWA), whereby *what is not known to be true* is always considered to be, simply, *not known.* Thus ontologies are particularly useful for dealing with domains, such as biological science, where our knowledge of the world is seen as being always incomplete, because the science itself is rapidly advancing.

The way that the closed-world assumption is practically made manifest on the standard database approach can be shown with the following simple example. We imagine a database consisting exactly of the information presented in this table:

Individuals	Attributes
Fido	dog
Rover	

An SQL query addressed to this database asking, "Is Rover an instance of a dog?" would yield the answer "no." On the open-world approach adopted by ontologists, in contrast, a query would not return *anything*, since we have no evidence as to whether Rover is or is not an instance of dog. Similarly, to the query "How many dogs are there?" the database would come back with an answer: exactly 1; the ontology would return: at least 1, but possibly more.

Third, and most important, a relational database schema exists primarily to constrain and structure the data, and it is very difficult to address complex queries over the data, or to use reasoners to check for consistency. The relations in a database themselves do not (and cannot) have properties inherent in them such as transitivity or symmetry. Rather, they are simply listed, like all other entries in the database. Given its basis in DL, (and thus in FOL), however, OWL's axioms and rules—including attributes of relations such as transitivity and symmetry—have been designed to facilitate the generation of implications and inferences from the ontologies that are formulated in its terms. For example, if you specify an OWL ontology in which the relation *is_pet_of* is defined to have domain *nonhuman animal* and range *person,* then if you assert the statement Rover *is_pet_of* Jim, you will be able conclude that Rover is a nonhuman animal and that Jim is a person.

OWL 2

Shortly after researchers started using OWL in the early 2000s, they began to realize its limitations. One basic problem with the initial version of OWL was its inability to express and reason over qualified cardinality restrictions. For example, while it is easy to say in OWL 1 that someone has four dogs as pets, it cannot be stated easily that someone has four dogs as pets, two of which are male; or that someone has four dogs as pets, one of which is a poodle and the other three are boxers. Another problem had to do with restrictions placed on literals. For example, while it is possible to say in OWL 1 that the barometric pressure in a particular part of the atmosphere has a value of 1,000 millibars, it is not possible to assert, for example, that this value is more than 900 and less than 1,100 millibars.[16] These and other problems prompted researchers to develop OWL 2, which became a W3C recommendation in October 2009.

OWL 2 also defines three profiles that are language subsets—EL, RL, and QL—each with useful computational properties and implementation capabilities. OWL 2 EL is ideal for large-scale ontologies, since using this profile allows many inference problems to be computed in polynomial time. OWL 2 RL is designed to take advantage of rule-based systems to solve inference problems. And in OWL 2 QL inference problems can be implemented as SQL queries against relational databases (RDBs), a very useful feature for practical, working ontologists.

Building Ontologies with Basic Formal Ontology

Tables 8.1 and 8.2 list some of the ontologies, institutions, and groups that have utilized BFO in developing their domain ontologies.[17] Having briefly described the basic features of Protégé, OWL, and associated resources, we are now in a better position to look at a few specific examples of domain ontologies that not only utilize BFO as upper-level ontology, but have also applied the principles and recommendations found in this book during ontology development.

Example: The Ontology for General Medical Science (OGMS)

An inspection of the homepage for the Ontology for General Medical Science (OGMS) reveals the influence of BFO and the recommendations from this book in the very definition of the ontology, which utilizes the Aristotelian definitional structure of "An A is a B that Cs":

The Ontology for General Medical Science (OGMS) is an ontology of entities involved in a clinical encounter. OGMS includes very general terms that are used across medical disciplines, including "disease," "disorder," "disease course," "diagnosis," "patient," and "healthcare provider." OGMS uses Basic Formal Ontology (BFO) as an upper-level ontology. The scope of OGMS is restricted to humans, but many terms can be applied to a variety of organisms. OGMS provides a formal theory of disease that can be further elaborated by specific disease ontologies.[18]

The domain of clinical medicine is a difficult one from an ontological perspective. Clinical terminology can be inconsistent, vague, and highly dependent on disciplinary context. OGMS is designed to provide a formal, explicit, nonredundant, and unambiguous representation of clinical terms that can begin to address these difficulties. OGMS is not a disease ontology; rather, it is a reference ontology that provides the terminological core of a general theory of disease and formal definitions for terms widely used in clinical encounters to describe different aspects of disease. It is being used as a framework for ontology modules for a range of different diseases and disease families.

OGMS utilizes the Aristotelian definitional structure for all of its entities, as in the example of constitutional genetic disorder, a disorder that pervades the whole organism, shown in figure 8.1, which is a Protégé-generated visualization of a fragment of OGMS that represents *constitutional genetic disorder* as a child of *genetic disorder*, itself a child of OGMS: *disorder*.

Figure 8.2 shows a fragment of the OWL representation of OGMS: *constitutional genetic disorder*. Notice the use of RDFS in line five, which reads

<rdfs:subClassOfrdf:resource="&obo;OGMS_0000047"/>

Because OGMS_0000047 is the alphanumeric identifier for "genetic disorder," this line communicates the fact that, in OGMS, "constitutional genetic disorder" is a subclass (subtype, kind) of "genetic disorder."

Figure 8.3 shows a portion of the Protégé-generated hierarchy for OGMS containing *constitutional genetic disorder* as part. It shows how the latter can be traced back in the ontology to BFO's *material entity* and from there to *independent continuant*. OGMS also utilizes the Relation Ontology.

Figure 8.4 shows how "disorder" fits into the broader context of OGMS, and the relations that connect it to the other core terms in the ontology, including terms that are among the most generally used in clinical medicine. What OGMS provides is a set of

Table 8.1
Ontologies utilizing BFO

ACGT Master Ontology	Alzheimer Disease Ontology	Adverse Event Ontology
Adverse Event Reporting Ontology	AFO Foundational Ontology	Actionable Intelligence Retrieval System
Bank Ontology	Beta Cell Genomics Application Ontology	BioAssay Ontology
Bioinformatics Web Service Ontology	Biological Collections Ontology	Biomedical Ethics Ontology
Biomedical Grid Terminology	BioTop	BIRNLex
Blood Ontology	Body Fluids Ontology	Cancer Cell Ontology
Cancer Chemoprevention Ontology	Cardiovascular Disease Ontology	Cell Behavior Ontology
Cell Cycle Ontology	Cell Expression, Localization, Development, and Anatomy Ontology	Cell Line Ontology
Cell Ontology	Chemical Entities of Biological Interest	CHRONIOUS Ontology Suite
Clusters of Orthologous Groups Analysis Ontology	Cognitive Paradigm Ontology	Common Anatomy Reference Ontology
Communication Standards Ontology	Conceptual Model Ontology	Coriell Cell Line Ontology
CPR Ontology	Document Act Ontology	Drug Interaction Ontology
Drug Ontology	Drug-drug Interaction Ontology	Earth Sciences Ontologies
Eagle-I Research Resource Ontology	Email Ontology	Emotion Ontology
Environment Ontology	Epidemiology Ontology	Evolution Ontology
Experimental Factor Ontology	Exposé	Financial Report Ontology
Flybase Drosophila Ontology	Fission Yeast Phenotype Ontology	Foundational Model of Anatomy
Gastrointestinal Ontology	Gene Regulation Ontology	General Information Model
Health Data Ontology Trunk	Human Interaction Network Ontology	Human Physiology Simulation Ontology
Infectious Disease Ontology	Information Artifact Ontology	Interdisciplinary Prostate Ontology Project
Lipid Ontology	Mental Disease Ontology	Mental Functioning Ontology
Middle Layer Ontology for Clinical Care	Military Ontology	MIRO and IRbase
Model for Clinical Information	Nanoparticle Ontology, Ontology for Cancer Nanotechnology Research	NeuroPsychological Testing Ontology
Neuroscience Information Framework	Neuroscience Information Framework Standard	Neural Electromagnetic Ontologies

Table 8.1 (continued)

Ocular Disease Ontology	OntoAlign++	Ontologized Information—BIobank
Ontology of Clinical Research	Ontology for Biomedical Investigations	Ontology for Drug Discovery Investigations
Ontology for General Medical Science	Ontology for Guiding Appropriate Antibiotic Prescribing	Ontology for Newborn Screening and Translational Research
Ontology for Mental Health and Quality of Life	Ontology of Biobanking Administration	Ontology for Parasite LifeCycle
Ontology for Rehabilitation (Traumatic Brain Injury)	Ontology of Data Mining	Ontology of Medically Related Social Entities
Ontology of Vaccine Adverse Event	Ontology-Based eXtensible Data Model	Oral Health and Disease Ontology
Parasite Experiment Ontology	Petrochemical Ontology	Phenotypic Quality Ontology
Plant Ontology	Population and Community Ontology	Proper Name Ontology
Protein-Ligand Interaction Ontology	Proteomics Data and Process Provenance Ontology	Protein Ontology
RNA Ontology	Saliva Ontology	Semantic HER
Senselab Ontology	Sequence Ontology	Sleep Domain Ontology
Semanticscience Integrated Ontology	Spatiotemporal Ontology for the Administrative Units of Switzerland	Special Nuclear Materials Detection Ontology
Subcellular Anatomy Ontology	Time Event Ontology	Translational Medicine Ontology
Universal Core Semantic Layer	Vaccine Ontology	Xenopus Anatomy Ontology
YAMATO	yOWL	Zebrafish Anatomical Ontology

coherent definitions for these terms, built around a view of a disease as a certain sort of power or potentiality—roughly, the potentiality for signs and symptoms to be manifested. The disease exists in the organism by virtue of physical disorders in that organism, for instance a disordered liver or a disordered lung. These powers or potentialities are BFO: *dispositions*, and so the OGMS defends the general view according to which diseases are special types of *disposition* that themselves can be traced up the *is_a* hierarchy to BFO: *specifically dependent continuant*, as shown in figure 8.5.

OGMS has been designed to serve as the framework for describing the entities involved in a clinical encounter. This means: describing how specific types of physical disorders relate to abnormal dispositions on the side of patients, dispositions realized

Table 8.2
Projects, institutions, and groups utilizing BFO

AstraZeneca—Clinical Information Science	Berkeley Bioinformatics Open-Source Projects	Biomedical Knowledge Engineering Lab at Seoul National University
Brain Operation Database	CTSAconnect	CUBRC
Data-Tactics Corporation	DOQS: Data-Oriented Quality Solutions	DSpace at NTNUA
Dumontier Lab	eagle-i Consortium	Elsevier Smart Content Strategy
EuPathDB	GoodOD	HIGHFLEET
U.S. Army Intelligence and Information Warfare Directorate	Influenza Research Database	INRIA Lorraine Research Unit
Kobe University Department of Sociomedical Informatics	Medical University Graz—Informatics, Statistics and Documentation	National Center for Multi-Source Information Fusion
National Center for Ontological Research	OBO (Open Biological and Biomedical Ontologies) Foundry	OntoCAT
OntoCOG	Open PHACTS	OpenEHR
REMINE	Saitama University	Science Commons—Neurocommons
Skeletome	University Hospital Erlangen—Radiology	University of Arkansas for Medical Sciences, Biomedical Informatics
University of Augsburg, Computer Science, Software Methodologies for Systems	University of Florida Biomedical Informatics	University of Texas Southwestern Medical Center
University of Washington Structural Informatics Group	Virus Pathogen Resource	VIVO

(manifesting themselves) in pathological processes that are recognized by the clinician in a clinical encounter as signs or symptoms and documented in clinical information systems. Clinical application ontologies extend OGMS by refining the basic taxonomic and relational structure for a particular domain of interest. Examples of such OGMS extensions include the Infectious Disease Ontology (IDO), Sleep Domain Ontology (SDO), Ontology of Medically Relevant Social Entities (OMRSE), Vital Sign Ontology (VSO), Mental Disease Ontology (OPMQoL), Neurological Disease Ontology (ND), Adverse Event Ontology (AEO), Ontology for Newborn Screening (ONSTR), Drug Ontology (DrOn), Model for Clinical Information (MCI), Ocular Disease Ontology (ODO), Oral Health and Disease Ontology (OHDO), Mental Functioning Ontology (MFO), and Cardiovascular Disease Ontology (CDO).[19]

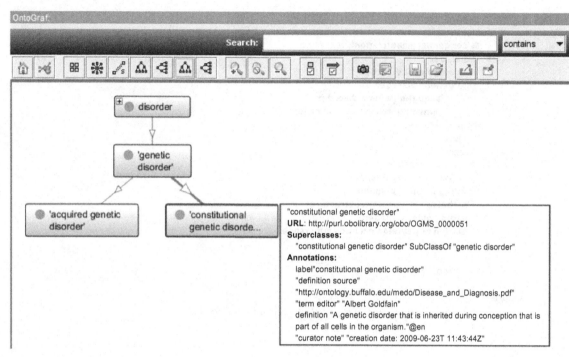

Figure 8.1

Protégé visualization of OGMS fragment for constitutional genetic disorder

```
<!-- http://purl.obolibrary.org/obo/OGMS_0000051 -->

<owl:Class rdf:about="&obo;OGMS_0000051">
  <rdfs:label
    >constitutional genetic disorder</rdfs:label>
  <rdfs:subClassOf rdf:resource="&obo;OGMS_0000047"/>
  <obo:IAO_0000117>Albert Goldfain</obo:IAO_0000117>
  <obo:IAO_0000119
    >http://ontology.buffalo.edu/medo/Disease_and_Diagnosis.pdf</obo:IAO_0000119>
  <obo:IAO_0000232
    >creation date: 2009-06-23T11:43:44z</obo:IAO_0000232>
  <obo:IAO_0000115 xml:lang="en"
    >A genetic disorder inherited during conception that is part of all cells in the organism.
    </obo:IAO_0000115>
</owl:Class>
```

Figure 8.2

Fragment of OWL representation of OGMS

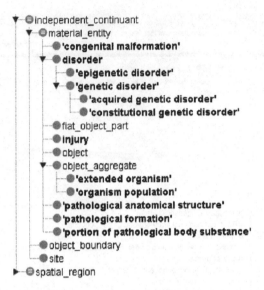

Figure 8.3
Constitutional genetic disorder as a type of independent continuant

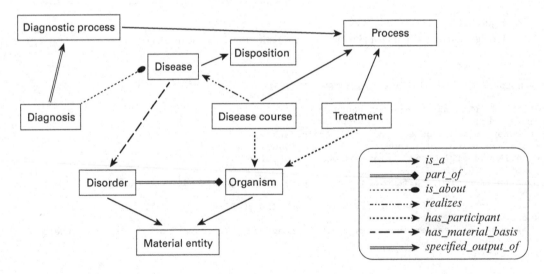

Figure 8.4
Selected core terms and relations of OGMS

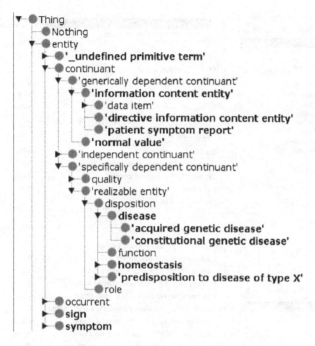

Figure 8.5
Disease as a subtype of specifically dependent continuant in OGMS

Infectious Disease Ontology (IDO)

The Infectious Disease Ontology (IDO) is an ontology representing over five hundred types or universals relevant to both the biomedical and clinical aspects of infectious diseases. A screenshot of IDO metadata generated by Protégé is shown in figure 8.6.

IDO consists strictly speaking of a suite of ontologies, each of which extends the general IDO-Core, which itself extends OGMS, which extends BFO in its turn. The extensions in the IDO suite are subdomain-specific ontologies primarily relating to specific infectious pathogens, such as IDO-Brucellosis, IDO-Dengue Fever, IDO-HIV, IDO-Infective Endocarditis, IDO-Influenza, IDO-Malaria, IDO-Staphylococcus Aureus, and IDO-Tuberculosis, as well as the vaccine ontology IDO-Vaccines.[20]

Figure 8.7 illustrates a fragment of the IDO in Protégé while figure 8.8 illustrates the position of the term "pathogen," defined as a material entity with a pathogenic disposition, in the IDO ontology. The circle next to this term in the left-hand column contains a tribar (≡) that indicates that the term is equivalent to some other term that behaves as a formal definition, in this case the following term:

material entity and has disposition some pathogenic disposition

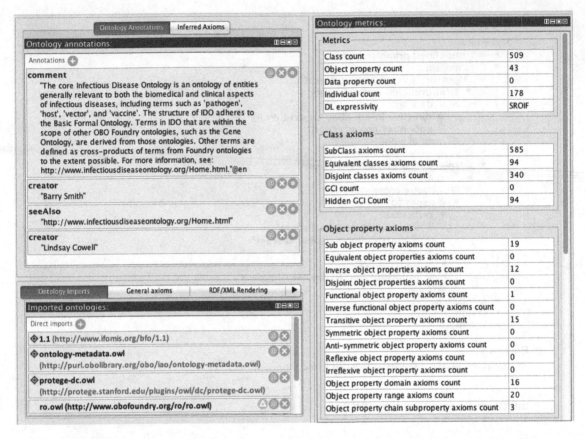

Figure 8.6
IDO metadata

Designating equivalency in this way is also a means by which ontologists may assert synonymy relations between terms in different ontologies.

It is good practice to reuse terms from one domain ontology in another domain ontology, and here, for example, the term "material entity" is taken over from BFO. IDO reuses many terms from other authoritative domain ontologies in formulating its definitions, including "disorder" from OGMS, "molecular entity" from the CHEBI (Chemical Entities of Biological Interest) ontology, and "macromolecular complex" from the Gene Ontology. An illustration of how IDO reuses terms from other ontologies is provided in figure 8.8, which illustrates the rendering of terms (using the Protégé View tab) by alphanumeric identifier.

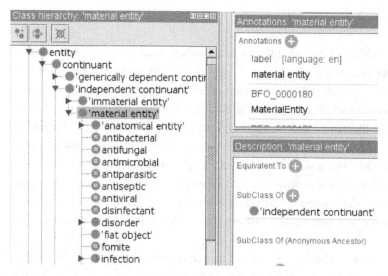

Figure 8.7
Protégé representation of a fragment of IDO

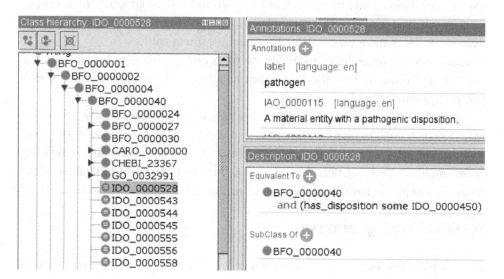

Figure 8.8
IDO representation of "pathogen" (IDO_0000528) and its definition

Information Artifact Ontology (IAO)

The Information Artifact Ontology (IAO) is an ontology of information entities based on BFO, emerging out of the need on the part of the developers of the Ontology for Biomedical Investigations (OBI) to categorize the different sorts of information entities that are involved in scientific research, including protocols, databases, experimental logs, published literature, and so forth.[21] IAO concerns the material bearers of information (books, hard drives, photographic prints, traffic signs), information content entities themselves (sentences in a book, XML files on a disk, symbols on a map, directions on a traffic sign), the processes that produce or consume information content entities (writing, documenting, encoding, drawing, client-server processing), and the relations between and among such entities (*is_about*, *denotes*, *is_translation_of*, and so forth).

A foundational idea in the IAO is that information content entities are related to other things by being "about" them or denoting them. Information content entities are a subtype of BFO's "generically dependent continuant." The ideas, tables, and figures being communicated right now in this book are examples of information content entities that denote other things such as the OWL and RDF languages, various ontologies, and entities of many other types. The hard copy of the book in your hand is an example of a material bearer of these information content entities.

Figure 8.9 illustrates IAO's treatment of the *scalar measurement datum*, which is defined, in conformity with the Aristotelian template, as a measurement datum with two parts: a numeral and a unit label.

Figure 8.10 shows one of IAO's data properties, *has_measurement_value*, which has IAO_0000032: *scalar measurement datum* as its domain, and as its range *float*. Thus *has_measurement_value* is a relation between a *scalar measurement datum* and a floating-point number. This relation is functional, meaning that an individual scalar measurement datum has one, and only one, measurement value of datatype float.

The Emotion Ontology (MFO-EM)

The Emotion Ontology (MFO-EM) is an extension of the Mental Functioning Ontology (MFO) covering mental processes (such as thinking) and dispositions (such as memory) in a BFO framework.[22] The Emotion Ontology itself comprises over 850 terms representing universals in the domain of affective phenomena such as emotions, moods, appraisals, and subjective feelings.[23] Each aspect of the ontology is rooted in BFO; for example, BFO: *occurrent* is utilized when defining *emotion occurrent* and its subtypes

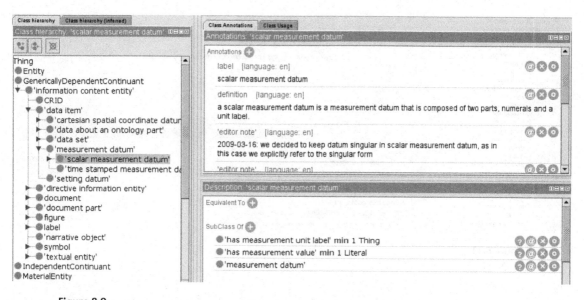

Figure 8.9

IAO: *scalar measurement datum*

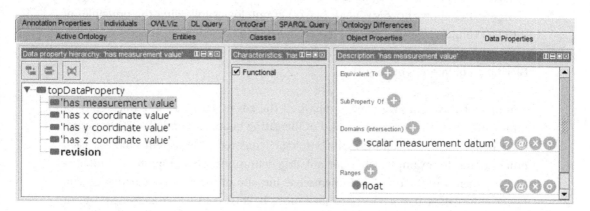

Figure 8.10

IAO's has_measurement_value data property

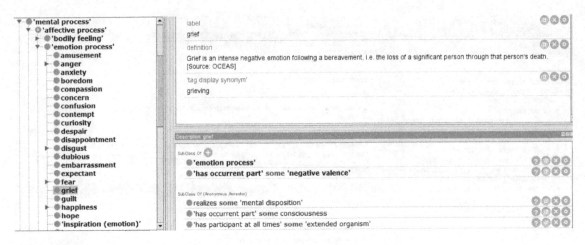

Figure 8.11
MFO-EM: *grief*

anger, happiness, and so forth. Similarly, emotion dispositions such as *love* and *hate* are classified in MFO-EM as subtypes of BFO: *disposition*.

Figure 8.11 provides a fragment of MFO-EM relating to the emotion occurrent: *grief.*

Figure 8.12 shows how MFO-EM, like other ontologies following the principles of the OBO Foundry, makes extensive use of the relations described in chapter 7.

Facilitation of Interoperability

Throughout this book, we have emphasized the advantages that interoperability of information systems brings to a world of increasing quantities of data. BFO promotes such interoperability by allowing the ontologies constructed in its terms to reuse each other's terms, for example, when formulating definitions. This promotes the integration not merely of the ontologies themselves but also of the respective bodies of data annotated in their terms. Queries that cannot be answered when addressed against single bodies of data often yield answers when data are combined through annotation with the same ontologies—so that multiple heterogeneous bodies of data behave as a unified target for ontology-based queries.

In chapter 5, we discussed the OBO Foundry initiative, "a collaborative experiment involving developers of science-based ontologies who are establishing a set of principles for ontology development with the goal of creating a suite of orthogonal interoperable reference ontologies in the biomedical domain."[24] Use of BFO and conformity to

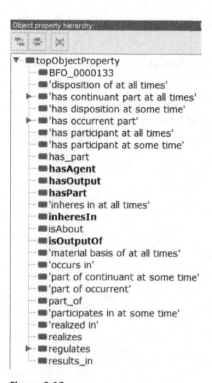

Figure 8.12
BFO relations used in MFO-EM

the principles found in this book are an integral part of the OBO Foundry approach, and are used also by OBO Foundry candidate ontologies such as OGMS, IAO, MFO, and MFO-EM, and many other ontologies. The result is an expanding virtual framework for navigating through massive amounts of biological and clinical data.

Further Reading on OWL, RDFS, and RDF

Baader, Franz, Diego Calvanese, Deborah L. McGuinness, Daniele Nardi, and Peter F. Patel-Schneider. *The Description Logic Handbook: Theory, Implementation and Applications.* Cambridge: Cambridge University Press, 2010.

Cimiano, Philipp, Christina Unger, and John McCrae. *Ontology-Based Interpretation of Natural Language.* San Rafael, CA: Morgan & Claypool, 2014.

Hitzler, Pascal, Markus Krötzsch, and Sebastian Rudolph. *Foundations of Semantic Web Technologies.* Boca Raton, FL: Chapman & Hall, 2009.

Horrocks, Ian. "Ontologies and the Semantic Web." *Communications of the ACM* 51 (12) (2008): 58–67.

Robinson, Peter N., and Sebastian Bauer. *Introduction to Bio-ontologies*. New York: Chapman and Hall/CRC, 2011.

Zhou, Yujiao, Bernardo Cuenca Grau, Ian Horrocks, Zhe Wu, and Jay Banerjee. "Making the Most of Your Triple Store: Query Answering in OWL 2 Using an RL Reasoner." In *Proceedings of the 22nd International Conference on World Wide Web (WWW 2013)*, ed. Ian Horrocks, 1569–1580. London: Elsevier, 2013.

Appendix on Implementation: Languages, Editors, Reasoners, Browsers, Tools for Reuse

Web Ontology Language (OWL)

As we saw, the Web Ontology Language (OWL) is a family of knowledge representation languages that are the basic computer-implementable languages in which ontologies are written. OWL languages are based on description logics. Whereas first-order logic is designed to capture the formal structure of sentences and to capture in great detail the inferential relations that obtain among them, description logics are designed to provide more information (material as well as formal) about the contents of propositions and concepts while expressing fewer of the inferential relations among them. The ability to express more information makes description logics more useful for knowledge representation and ontology purposes, while the limited inferential information that they represent makes them more tractable for computers.

Editors: Protégé and OBO Editor

Because OWL programing languages and description logics are somewhat complicated to interact with directly, ontology authoring and editing software has been developed. The most widely used tool of this sort is Protégé, which uses OWL and is implemented in the Java programming language. Rather than requiring users to be intimately familiar with the syntax and commands of some particular version of OWL, Protégé provides a uniform ontology viewing and editing experience that allows users to select commands, relations, and so on, from drop boxes and to enter definitions and other information in ordinary language. This software is freely available, along with tutorials and other user information, at http://protege.stanford.edu/. Another significant ontology editor is OBO-Edit, which is optimized for reading and writing ontologies in the OBO biological ontology file format. More information about OBO-Edit is available at http://oboedit.org.

Reasoners

Semantic reasoners (or simply *reasoners*) are pieces of software designed to infer the consequences of assertions in an ontology. Once an ontology has been implemented in some OWL language—for example, through an editor such as Protégé or OBO-Edit— much of the information that has been entered in the ontology has an explicit structure that can be understood as an axiom or assertion. Reasoners use logical rules to draw conclusions from such axioms or assertions. This means that from the assertions in an ontology a reasoning program can determine new relations between universals or classes in the ontology (for instance, by tracing *is_a* or *part_of* relations through the ontology based on their transitivity). Such reasoners can be used to check for consistency of the information entered into the ontology. Reasoners in standard use with OWL include: Pellet (http://clarkparsia.com/pellet/), FaCT++ (http://owl.man.ac.uk/factplusplus/), and HermiT Owl Reasoner (http://hermit-reasoner.com/).

Browsers

Ontology browsers are tools that allow a user to visualize and explore single ontologies and also to query and compare a number of ontologies simultaneously. A very comprehensive ontology browser for biological and biomedical purposes is Bioportal (http://bioportal.bioontology.org), which allows users to search for relevant terms across and to browse in a very large and ever-expanding repository of existing ontologies. Another is the University of Michigan Medical School's Ontobee (http://www.ontobee.org/), which is the default linked ontology data server for OBO Foundry library ontologies. As noted on its website, Ontobee is "aimed to facilitate ontology data sharing, visualization, query, integration, and analysis. Ontobee dynamically dereferences and presents individual ontology term URIs to (i) *HTML web pages* for user-friendly web browsing and navigation, and to (ii) *RDF source code* for Semantic Web applications."[1] Ontology browsers are important for purposes of ontology design because they allow would-be designers to survey existing ontologies so as to avoid reinventing the wheel. By making it possible to compare the entry for a single term across multiple ontologies, they also make it possible to do quality control, checking to see which ontology has the best or most scientifically accurate information. Other browsers include QuickGo, a browser for Gene Ontology terms and annotations (http://www.ebi.ac.uk/QuickGO/), and the Ontology Lookup Service (http://www.ebi.ac.uk/ontology-lookup/).

Tools for Reuse: Ontofox and Mireot

While ontology browsers make the task of searching for and reusing existing ontologies easier, by themselves they still leave a potential user with the task of finding some way to copy the relevant data from existing ontologies into their own ontologies. In response to this problem, researchers have begun developing software the specific goal of which is to incorporate parts of existing ontologies (perhaps discovered via an ontology browser) for use elsewhere. For example, OntoFox is a project driven by the principles of MIREOT (minimum information to reference an external ontology term), which have as their aim making it both possible and efficient for application ontology designers to utilize preexisting reference ontologies in their application designs rather than creating new content for each new application.[2] Also, the OntoFox tool (http://ontofox.hegroup.org/introduction.php) can retrieve selected portions of taxonomies from already-existing ontologies based on a specification of the highest (most general) and lowest (most specific) terms from some section of the ontology that are of interest to the searcher.

Glossary

Adequatism.
The view that the entities in any given domain should be taken seriously on their own terms (as contrasted with **reductionism**). The goal of adequatism is to do justice to all the different kinds of entities that exist in reality; see also **General principles of ontology design**.

All-some structure.
The structure that applies to the relations obtaining between **universals** whereby if a universal C bears some relation R to a universal D, then all relevant instances of C must bear the relevant instance-level relation to some instance of D at all relevant times.

Aristotelian definitional structure.
A definition that has the basic form "An A is a B that Cs" where "A" is the *definiendum*, the term that is being defined; "B that Cs" is the *definiens*, the expression that does the defining; and "C" is a statement of the *differentia*, that is to say, a statement of what it is that marks out those instances of B that are As (those features Bs must possess in order to be As).

Basic Formal Ontology (BFO).
A top-level (or upper-level) ontology consisting of **continuants** and **occurrents** developed to support integration especially of data obtained through scientific research. BFO is deliberately designed to be very small, in order to represent in consistent fashion those top-level categories common to domain ontologies developed by scientists in different fields. BFO assists domain ontologists by providing a common top-level structure to support the interoperability of the multiple domain ontologies created in its terms.

Category.
A formal (which is to say, domain-neutral) **universal**, such as *entity, continuant, occurrent*.

Class.
A maximal collection of particulars falling under a given general **term**; also called the extension of the term (or of the **universal** that the term denotes).

Continuant.
An entity that *continues* or *persists* through time, including (1) independent objects, (2) qualities and dispositions, and (3) the spatial regions these entities occupy at any given time. **Continuant** and **occurrent** are the two highest categories (universals) in BFO.

Continuant fiat boundary.
An **immaterial entity** that is of zero, one, or two dimensions and does not include a spatial region as part. Intuitively, a *continuant fiat boundary* is a boundary of some material entity that exists exactly where that object meets its surroundings.

Controlled vocabulary.
A collection of preferred terms that are used to promote consistent description and retrieval of data.

Defined class.
A collection of **individuals** that are grouped together by virtue of their exhibiting some combination of characteristics that does not correspond to any **universal**.

Description logic (DL).
A fragment of **first-order logic** (FOL) used for purposes of formal knowledge representation and having more efficient decision properties than FOL. See also **Web Ontology Language (OWL)**.

Disposition.
A **realizable entity** (a power, potential, or tendency) that exists because of certain features of the physical makeup of the **independent continuant** that is its bearer.

Domain.
A delineated portion or sphere of reality corresponding to a scientific discipline (such as cell biology or electron microscopy), or to an area of knowledge or interest such as the Great War or stamp collecting.

Domain ontology.
See **Ontology, domain**.

Entity.
Anything that exists.

Epistemological idealism.
See **Idealism**.

Fallibilism.
The view that, although our current scientific theories are the best candidates we have for representing the truth about reality, it may nevertheless be the case at any given stage that elements of our best current knowledge are incorrect; see **General principles of ontology design**.

Fiat object part.
A **material entity** that is a proper part of some larger object, but is not demarcated from the remainder of this object by any physical discontinuities (thus, it is not itself an object). Examples include your upper torso, the Western hemisphere of the planet earth.

First-order logic (FOL).
A formal language and system of reasoning utilizing predicates, constants and variables, quantifiers (all, some, none), and logical connectives (and, not, or, material implication); also known as first-order predicate logic; see also **Description logic (DL).**

Foundational relations.
The fundamental relations in BFO, especially *is_a* (meaning "is a subtype of") and relations of part to whole.

Function.
A **realizable entity** whose realization is an end- or goal-directed activity of its bearer that exists in the bearer in virtue of its having a certain physical makeup as a result of either natural selection (in the case of biological entities) or intentional design (in the case of artifacts).

General principles of ontology design.
The principles to be applied in every process of designing an ontology, including **realism, perspectivalism, adequatism, fallibilism,** the **open-world assumption,** and the **low-hanging fruit principle.**

Generically dependent continuant.
A **continuant** that is dependent on one or other **independent continuants** and can migrate from one bearer to another through a process of copying. We can think of generically dependent continuants as complex continuant patterns either of the sort created by authors or designers or (in the case of DNA sequences) brought into being through the processes of evolution.

Granularity.
The distinction between levels of reality exemplified for example in the biological domain in the levels of cells, organs, organisms and populations. See **Perspectivalism.**

History.
A BFO: *process* that is the sum of the totality of processes taking place in the spatiotemporal region occupied by a material entity or site.

Idealism.
The thesis that our perceptions, thoughts, and statements are not about reality, but are rather about certain mental or created objects (variously called appearances, concepts, ideas, or models). For the idealist, there is nothing that exists (or, for the defender of **Epistemological idealism,** nothing that can be known) outside of the realm of sensations or ideas or concepts; see also **Realism.**

Immaterial entity.
An **independent continuant** that contains no **material entities** as parts. Immaterial entities divide into two major subgroups: (1) boundaries and sites, which bound, or are demarcated in relation to, material entities, and which can thus change location, shape, and size as their material hosts move or change shape or size; (2) spatial regions, which exist independently of material entities, and which thus do not change.

Independent continuant.
A **continuant** entity that is the bearer of qualities and a participant in processes. Independent continuants are such that their identity can be maintained over time through gain and loss of parts, as well as through changes in qualities.

Individual.
See **Particular.**

Inherence.
A one-sided dependence that obtains between **specifically** and **generically dependent continuants** on the one hand and **independent continuants** on the other. **Qualities, dispositions,** and **roles** inhere in independent continuants.

Instance.
See **Particular.**

Instantiation.
A relation between a **particular** and a **universal**; the particular is one of an open-ended set of particulars that are similar in the relevant respect; the particular is such that if it ceases to be an instance of this universal then it ceases to exist.

Inventory.
A representational artifact consisting of entries designed to keep track of the particulars contained, for example, in a warehouse of products.

is_a.
The subtype relation used to form the backbone taxonomic hierarchy of an ontology.

Low-hanging fruit principle.
The principle that states that the designer of a domain ontology should start with those (often trivial) features of the subject matter that are the easiest and clearest to understand; also see **General principles of ontology design.**

Material entity.
An **independent continuant** that has some portion of matter as part, is spatially extended in three dimensions, and that continues to exist through some interval of time, however short. Three principal subtypes of material entity are **object, fiat object part,** and **object aggregate.**

Nominalism.
The position that there are no **universals**, in contrast to **ontological realism**; also see **Representationalism**.

Object.
A **material entity** that is (1) spatially extended in three dimensions; (2) causally unified; and (3) maximally self-connected. Examples include a single cell, a laptop, an organism, a planet, a spaceship.

Object aggregate.
A **material entity** that has as parts (exactly) two or more **objects** that are separate from each other in the sense that they share no parts in common. Examples include a heap of stones, a population of bacteria, a flock of geese.

Occurrent.
An entity that unfolds itself in time; the BFO category of *occurrents* comprises not only (1) the **processes** that unfold themselves in their successive temporal phases, but also (2) the boundaries or thresholds at the beginnings or ends of such temporal phases, as well as (3) the **temporal** and **spatio-temporal regions** in which these processes occur. **Occurrent** and **continuant** are the two highest categories (universals) in BFO.

One-dimensional continuant fiat boundary.
A **continuant fiat boundary** that is a continuous fiat line whose location is defined in relation to some **material entity**. Examples include the boundary of a real estate parcel, your waistline.

One-dimensional spatial region.
A **spatial region** with one dimension, also called a spatial line, defined relative to some reference frame; for example, lines of latitude and longitude.

One-dimensional temporal region.
A **temporal region** that is extended in time. It has further temporal regions as parts. One-dimensional temporal regions are the temporal regions in which processes occur or unfold.

Ontological realism.
The view according to which the general truths about reality that science discovers are grounded in universals, which are common features and characteristics of the entities in reality in virtue of which they are grouped into circles of similars; compare **Nominalism**, **Representationalism**.

Ontology.
A **representational artifact**, comprising a taxonomy as proper part, whose representations are intended to designate some combination of universals, defined classes, and certain relations between them; see also **Ontology, philosophical**.

Ontology, application.
An application ontology is an ontology created to accomplish some task of local significance. See **Ontology, reference.**

Ontology, domain.
A domain ontology is an ontology that describes and categorizes some **domain.**

Ontology, formal.
The study of the universals, relations, and structures common to all domains of reality; also used to refer to an upper-level ontology such as BFO.

Ontology, material.
The study of the universals, relations, and structures common to some specific domain of reality; sometimes used as a synonym of "**domain ontology.**"

Ontology, philosophical.
The theory of what exists—the study of the kinds of entities in reality and of the relationships that these entities bear to one another; also known as *metaphysics*. The goal of philosophical ontology is to provide a clear, coherent, and rigorously worked out account of the basic structures of reality.

Ontology, reference.
An ontology that is intended to provide a comprehensive representation of the entities in a given domain encapsulating the terminological content of established knowledge of the sort that is contained in a scientific textbook.

Ontology, top- (or upper-) level.
An ontology of highly general categories and relations that subsume the universals represented by specific domain ontologies.

Open-world assumption.
The assumption that we capture knowledge within an ontology or ontology-like resource in an ongoing process as we discover it, so that we can at no stage guarantee that we have discovered complete information—hence no conclusions should be drawn from the fact that a given assertion is not recorded in our system; see **General principles of ontology design.**

Participation.
The relation between a **material entity** and a **process** that obtains in virtue of the fact that the former participates in the latter.

Particular.
An individual (nonrepeatable) denizen of reality (an instance of a **universal**); all particulars stand in the relation of *instantiates* to some universal; each particular occupies a unique spatiotemporal location.

Particular-particular relation.
A relation between one particular and another; also called an instance-level relation; for example: Mary's leg **part_of** Mary; see **Relation.**

Perspectivalism.
The view according to which reality is too complex and variegated to be embraced within a single scientific theory, and that multiple distinct scientific theories may be equally accurate representations of one and the same reality, for instance because they partition this reality at different levels of **granularity**; see **General principles of ontology design**.

Principle of reuse.
A principle stating that ontologies should as far as possible reuse the ontological content that has already been created; see **General principles of ontology design**.

Process.
An **occurrent** entity that exists in time by occurring or happening, has temporal parts, and always depends on at least one **independent continuant** as participant.

Process boundary.
An **occurrent** entity that is the instantaneous temporal boundary of a **process**.

Quality.
A **specifically dependent continuant** that, if it inheres in an entity at all, is fully exhibited, manifested, or realized in that entity. In order for a quality to exist, one or more independent continuants must also exist. Examples include the mass of a kidney, the color of this portion of blood, and the shape of a hand.

Realism.
The view that thought, experience, and knowledge are (if partially and fallibly) about reality. A view of this sort should be the general attitude to be kept in mind throughout the process of designing an ontology; see **General principles of ontology design**.

Realizable entity.
A **specifically dependent continuant entity** that has at least one **independent continuant** as its bearer, and whose instances can be realized (manifested, actualized, executed) in associated processes of specific correlated types in which the bearer participates.

Reflexivity.
The property of a relation R whereby anything that bears R to something also bears that relation to itself. The relation "is as tall as" is reflexive, because when John is as tall as Jill, then he also stands in this same relation to himself.

Relation.
The manner in which two or more entities are associated or connected together. BFO recognizes three basic types of relation: connecting **universal** to **universal**, **universal** to **particular**, and **particular** to **particular**.

Relational quality.
A **quality** that inheres in two or more independent continuants. Examples include a marriage bond, a debt, an agreement. From the BFO perspective there is both the relational quality **universal** *marriage bond* as well as specific instantiations of this universal obtaining between (and specifically depending upon) John and Mary, Bill and Sally, and so forth.

Representation.
An entity that makes reference to or is about another entity or entities.

Representational artifacts.
A **representation** that has been produced by someone and made available in a form that allows it to be accessed by others (such as a drawing, map, book, or computer database).

Representationalism.
The view that our perceptions, thoughts, beliefs, or models are directly about concepts (or ideas or images) in our minds and only indirectly about nonmental entities of various sorts in reality. On this view, what we actually know are not things in reality, but the ways in which we experience and conceptualize these things; see **Idealism**, **Realism**.

Representationalist interpretation of knowledge.
The view that the goal of knowledge representation is to represent concepts or ideas; see **Representationalism**, **Idealism**, **Realism**.

Role.
A **realizable entity** that (1) exists because the bearer is in some special physical, social, or institutional set of circumstances in which the bearer does not have to be, and (2) is not such that, if this realizable entity ceases to exist, then the physical make-up of the bearer is thereby changed. A role is thus always optional.

Semantic interoperability.
A property that obtains between two (or more) data or information systems when they are such that, because their terms are defined according to a common, logically well-defined ontology, each can carry out the tasks for which it was designed using data and information taken from the other as seamlessly as it can when using its own data and information.

Semantic Web.
Whereas the World Wide Web (WWW) is an interconnected system of web pages, the Semantic Web (SW) is an interconnected system made up of the content (data and information) from those pages. The SW emerged out of thinking in the field of artificial intelligence and was conceived as a system that enables machines to "understand" and respond to complex human queries based on their meaning—hence the use of the term "semantic."

Site.
An **immaterial entity** in which objects such as molecules of air or organisms can be contained.

Spatial region.
A **continuant entity** that is a part of space. When an object moves from one place to another, it occupies a continuous series of different three-dimensional spatial regions at different times.

Spatiotemporal region.
An **occurrent** entity at or in which **occurrent** entities can be located. Just as the continuant representation of entities views space as a container within which objects and their qualities exist, so too the occurrent representation of processes views the combination of space and time together as such a container, within which processes unfold.

Specifically dependent continuant.
A **continuant** entity that depends on precisely one independent continuant for its existence. The former is dependent on the latter in the sense that, if the latter ceases to exist, then the former will as a matter of necessity cease to exist also. See **Independent continuant, Generically dependent continuant.**

Symmetry.
The property of a relation R whereby if a thing A bears R to something else B, then B also bears R to A. Example: if A is adjacent to B, then B is also adjacent to A.

Taxonomy.
A representational artifact taking the form of a graph with nodes representing kinds of things (universals) and edges representing subtype or subclass (*is_a*) relations among these types of things. The most familiar kind of taxonomy is the classification of living things: domain, kingdom, phylum, class, order, family, genus, and species.

Temporal region.
An **occurrent** entity that is a part of time.

Term.
A noun or noun phrase, understood as a linguistic sign, that is utilized to represent some entity in the world.

Terminology.
A **representational artifact** containing a list of terms, complete with definitions, used in some domain and formulated in a natural language.

Three-dimensional spatial region.
A **spatial region** with three dimensions, also called a spatial volume; for example, the region occupied at any given time by the planet earth.

Top-level ontology.
See **Ontology, top-level.**

Transitivity.
The property of a relation R whereby if a thing A bears R to another thing B, and B bears R to some third thing C, then A also bears R to C. A simple example is *being taller than*.

Two-dimensional continuant fiat boundary.
A **continuant fiat boundary** that is a self-connected fiat surface whose location is defined in relation to some material entity; for example, any surface of a continuant material object separating that object from the rest of its environment.

Two-dimensional spatial region.
A **spatial region** with two dimensions, also called a spatial surface; for example, the region occupied by the surface of the earth.

Universal.
A mind-independent, repeatable feature of reality that exists only as instantiated in a respective **particular** (individual thing, instance) and is also dependent upon a particular for its existence. For example, the two universals *red* and *ball* are instantiated in a red ball lying on the floor. All particulars stand in the instantiation relation to some universal. Universals are the sorts of entities that are represented by general terms used in the formulation of scientific laws.

Use-mention distinction.
The distinction between *using* a noun phrase to make reference to something in reality, and *mentioning* the same noun phrase in order to engage in discourse about this noun phrase itself.

Web Ontology Language (OWL).
A family of languages used by the Semantic Web.

Zero-dimensional continuant fiat boundary.
A **continuant fiat boundary** that is a fiat point whose location is defined in relation to some material entity. Examples include the North Pole and the earth's center of gravity.

Zero-dimensional spatial region.
A **spatial region** with no dimensions, also called a spatial point; for example, the point of origin of a spatial coordinate system.

Zero-dimensional temporal region.
A **temporal region** that is without extent. Zero-dimensional temporal regions, also called temporal instants, are the temporal regions in which process boundaries are located.

Web Links Mentioned in the Text Including Ontologies, Research Groups, Software, and Reasoning Tools

Basic Formal Ontology (BFO)	http://ifomis.org/bfo
Basic Formal Ontology Discussion Group	http://groups.google.com/group/bfo-discuss
Cell Ontology (CL)	http://obofoundry.org/cgi-bin/detail.cgi?id=cell
Chemical Entities of Biological Interest (ChEBI)	http://www.ebi.ac.uk/chebi/
Descriptive Ontology for Linguistic and Cognitive Engineering (DOLCE)	http://www.loa.istc.cnr.it/old/DOLCE.html
Disease Ontology (DO)	http://disease-ontology.org
Emotion Ontology (EO)	http://www.ontobee.org/browser/index.php?o=MFOEM
Foundational Model of Anatomy (FMA)	http://sig.biostr.washington.edu/projects/fm/
Gene Ontology (GO) Consortium	http://www.geneontology.org/
Gramene Plant Environment Ontology	http://archive.gramene.org/db/ontology/search?id=EO:0007359
Health Level 7 (HL7)	http://www.hl7.org/
HermiT OWL Reasoner	http://hermit-reasoner.com/
The Human Phenotype Ontology (HPO)	http://www.human-phenotype-ontology.org
Infectious Disease Ontology (IDO)	http://www.obofoundry.org/cgi-bin/detail.cgi?id=infectious_disease_ontology
Information Artifact Ontology (IAO)	http://www.ontobee.org/browser/index.php?o=IAO
Mental Functioning Ontology (MFO)	http://www.ontobee.org/browser/index.php?o=MF
The ImmuneXpresso project	http://www.immport-labs.org/
International Classification of Diseases from the World Health Organization	http://www.who.int/classifications/icd/en/
Laboratory for Applied Ontology	http://www. loa-cnr.it/Location.html
Medical Subject Headings (MeSH) Database	http://umlsks.nlm.nih.gov/
National Cancer Institute Thesaurus (NCIT)	http://www.nci.nih.gov/cancerinfo/terminologyresources

National Center for Biotechnology Information (NCBI)	http://www.ncbi.nlm.nih.gov/
NIH Neuroscience Information Framework	http://www.neuinfo.org/
OBO Relation Ontology (RO)	http://obofoundry.org/ro/
OntoFox Introduction	http://ontofox.hegroup.org/introduction.php
Ontology for Biomedical Investigations (OBI)	http://obi.sourceforge.net/
Ontology for General Medical Science (OGMS)	http://www.obofoundry.org/cgi-bin/detail .cgi?id=OGMS
Open Biomedical Ontologies (OBO) Consortium	http://www.obofoundry.org/
Pellet Reasoner	http://clarkparsia.com/pellet/protege/
Phenotypic Quality Ontology (PATO)	http://www.obofoundry.org/cgi-bin/detail .cgi?id=quality
Protein Ontology (PRO)	http://proteinontology.info/
Protégé Ontology Editor	http://protege.stanford.edu/
RacerPro Reasoner	http://franz.com/agraph/racer/
Sequence Ontology (SO)	http://www.sequenceontology.org/
Situational Awareness and Preparedness for Public Health Incidents Using Reasoning Engines (SAPPHIRE)	http://www.w3.org/2001/sw/sweo/public/ UseCases/UniTexas/
Standard Upper Merged Ontology (SUMO)	http://suo.ieee.org/
Unified Medical Language System (UMLS)	http://www.nlm.nih.gov/research/umls/
W3C Web Ontology Language (OWL)	http://www.w3.org/TR/owl2-overview/

Notes

Introduction

1. Mik Miliard, "Data Variety Bigger Hurdle than Volume," *HealthcareITNews*, July 3, 2014, http://www.healthcareitnews.com/news/data-variety-bigger-hurdle-volume?topic=02,06&mkt_tok=3RkMMJWWfF9wsRonuq3IZKXonjHpfsX87OQkWbHr08Yy0EZ5VunJEUWy2YIDT9Q%2FcOedCQkZHblFnVUKSK2vULcNqKwP, accessed August 25, 2014.

2. Karen B. DeSalvo and Erica Galvez, "Connecting Health and Care for the Nation: A 10-Year Vision to Achieve an Interoperable Health IT Infrastructure," The Office of the National Coordinator for Health Information Technology, 2014, p. 4, http://www.healthit.gov/sites/default/files/ONC10yearInteroperabilityConceptPaper.pdf, accessed September 1, 2014.

3. Greg Slabodkin, "EHR Interoperability Key to Modernizing Clinical Trials," *HealthData Management*, July 10, 2014, http://www.healthdatamanagement.com/news/EHR-Interoperability-Needed-to-Improve-Clinical-Trials-48392-1.html, accessed August 25, 2014.

4. Susan J. Grobe, "ICNP Version 1: International Classification for Nursing Practice—A Unified Nursing Language System," 2005, www.nicecomputing.ch/nieurope/S%20Grobe%20ICNP.pdf, accessed August 30, 2014.

5. U.S. Department of Health and Human Services, "Development of Software and Analysis Methods for Biomedical Big Data in Targeted Areas of High Need (U01)," 2014, http://grants.nih.gov/grants/guide/rfa-files/RFA-HG-14-020.html, accessed August 25, 2014.

6. This definition is taken over from the HL7 Reference Information Model (RIM) Version V 02-07, as described in section 3.2.5, http://www.vico.org/CDAR22005_HL7SP/infrastructure/rim/rim.htm, accessed July 25, 2014.

7. In a later version of its documentation HL7 corrects this second error. It now defines a *living subject* as "anything that essentially has the property of life, independent of current state (a dead human corpse is still essentially a living subject)." See Health informatics, HL7 version 3, Reference information model, Release 4, Document ISO/HL7 21731:2011(E), http://www.hl7.org/index.cfm. We note that the first error still remains.

8. BRIDG Version 1.0. Phase 1.0. Created on January 5, 2005; last modified December 14, 2006. All releases are available at http://bridgmodel.nci.nih.gov/, accessed August 25, 2014.

9. Compare also the case of Microsoft HealthVault, which defines "an allergy episode [as] a single unit of data that is recorded in Microsoft HealthVault," last updated 2014, http://msdn .microsoft.com/en-us/library/aa155110.aspx, accessed August 25, 2014.

10. BRIDG Version 3.2, created May 9, 2014, http://bridgmodel.nci.nih.gov, accessed August 25, 2014.

11. Both examples taken from the Fast Healthcare Interoperability Resources Specification (FHIR), http://www.hl7.org/implement/standards/fhir/v3/EntityClass/index.html, accessed June 6, 2014.

12. Health Level 7 FHIR Development Version, http://hl7.org/implement/standards/FHIR-Develop/v3/RoleClass/index.html, accessed September 29, 2014.

1   What Is an Ontology?

1. Taken from Barry Smith, Waclaw Kusnierczyk, Daniel Schober, and Werner Ceusters, "Towards a Reference Terminology for Ontology Research and Development in the Biomedical Domain," in *Proceedings of the 2nd International Workshop on Formal Biomedical Knowledge Representation* (KR-MED 2006), vol. 222, ed. Olivier Bodenreider (Baltimore, MD: KR-MED Publications, 2006), 57–66, http://www.informatik.uni-trier.de/~ley/db/conf/krmed/krmed2006.html, accessed December 17, 2014.

2. See http://bioportal.bioontology.org/ontologies/MF, accessed August 4, 2014.

3. ISO 1087–1:2000, Terminology Work—Vocabulary—Part 1: Theory and Application, 2000.

4. Ibid.

5. Barry Smith, Werner Ceusters, and Rita Temmerman, "Wüsteria," *Studies in Health Technology and Information* 116 (2005): 647–652.

6. Christopher G. Chute, "Medical Concept Representation," in *Medical Informatics: Integrated Series in Information Systems*, vol. 8, ed. H. Chen, S. S. Fuller, C. Friedman, and W. Hersh (New York: Springer, 2005), 163–182. James J. Cimino, "In Defense of the Desiderata," *Journal of Biomedical Informatics* 39, no. 3 (2006): 299–306.

7. Stefan Schulz et al., "From Concept Representations to Ontologies: A Paradigm Shift in Health Informatics?" *Healthcare Informatics Research* 19, no. 4 (2013): 235–242.

8. See Ronald J. Brachman and Hector J. Levesque, eds., *Readings in Knowledge Representation* (San Francisco: Morgan Kaufmann Publishers Inc., 1985).

9. And also common-sense beliefs, as in J. R. Hobbs and R. C. Moore, eds., *Formal Theories of the Common-Sense World* (Norwood, NJ: Ablex, 1985).

10. G. Van Heijst, A. T. Schreiber, and B. J. Wielinga, "Using Explicit Ontologies in KBS Development," *International Journal of Human–Computer Studies* 45 (1996): 183.

11. See his "A Translation Approach to Portable Ontologies," *Knowledge Acquisition* 5, no. 2 (1992): 199–220. Note that for Gruber himself—though not for many of those who follow in his wake—conceptualizations are to be conceived not as creatures of the mind, but rather as artifacts analogous to software programs.

12. For the survey results and discussion, see David Bourget and David Chalmers, eds., "The PhilPapers Survey," *PhilPapers*, n.d., http://philpapers.org/surveys/, accessed August 15, 2014.

13. James Franklin, "Stove's Discovery of the Worst Argument in the World," *Philosophy* 77 (2002): 615–624.

14. "SNOMED CT," http://www.ihtsdo.org/snomed-ct/, accessed August 4, 2014.

15. International Health Terminology Standards Development Organisation, *SNOMED CT® Technical Reference Guide—July 2010 International Release* (Washington, DC: College of American Pathologists, 2010).

16. "In computer science the expressions up for semantic evaluation do in fact refer very often to things inside the computer—to subroutines that can be called, to memory addresses, to data structures, etc." Daniel Dennett, *Brainchildren: Essays on Designing Minds* (Cambridge, MA: MIT Press, 1998), 281.

17. See http://lists.hl7.org/read/messages?id=111079, accessed August 15, 2014. This message is no longer accessible at the HL7 site, but is archived here: http://hl7-watch.blogspot.com/2007/09/piece-of-good-news-has-been-posted-on.html.

18. "HealthRecordItem Class," http://msdn.microsoft.com/en-us/library/microsoft.health.health recorditem.aspx, accessed August 4, 2014.

19. "Allergy Class," http://msdn.microsoft.com/en-us/library/microsoft.health.itemtypes.allergy .aspx, accessed August 4, 2014.

20. We use "instance" and "particular" as synonyms, the former term being used where we wish to draw out the relation of instantiation between a universal and its particular instances.

21. For further background, see Barry Smith and Werner Ceusters, "Ontological Realism: A Methodology for Coordinated Evolution of Scientific Ontologies," *Applied Ontology* 5, nos. 3–4 (2010): 139–188.

22. Franklin, "Stove's Discovery."

23. A somewhat less radical version of this view, called "resemblance nominalism," holds that some things are in some sense objectively similar to other things and thereby form circles of similars, with which our general concepts or general words are associated. See, for example, G. Rodriguez-Pereyra, *Resemblance Nominalism: A Solution to the Problem of Universals* (Oxford: Clarendon Press, 2002).

24. See Barry Smith and Werner Ceusters, "Strategies for Referent Tracking in Electronic Health Records," *Journal of Biomedical Informatics* 39, no. 3 (June 2006): 362–378.

25. For purposes of illustration we here treat biological species as though they are themselves universals in the sense described in the text. This is one view of the matter accepted by many biologists. However in contemporary philosophy of biology species are often viewed as complex particular entities consisting of whole current and historical populations. For our purposes here it is not necessary to take a stand on the matter. For a thorough overview of the issues, see Marc Ereshefsky, "Species," *The Stanford Encyclopedia of Philosophy* (Spring 2010 edition), ed. Edward N. Zalta, http://plato.stanford.edu/archives/spr2010/entries/species/, accessed August 5, 2014.

2 Kinds of Ontologies and the Role of Taxonomies

1. World Health Organization, "International Classification of Diseases (ICD)," http://www.who.int/classifications/icd/en/, accessed August 5, 2014.

2. U.S. Army, "Joint Doctrine Hierarchy," n.d., http://usacac.army.mil/cac2/doctrine/CDM pages/cdm_joint heirarchy.html, accessed August 5, 2014.

3. Defined classes may also stand in the *is_a* relation to each other, as we will briefly discuss in chapter 7, where a more careful definition of *is_a* is given.

4. H-S. Low, C. J. O. Baker, A. Garcia, and M. R. Wenk, "An OWL-DL Ontology for Classification of Lipids," in *Proceedings of the International Conference on Biomedical Ontology* (ICBO 2009) (Buffalo, NY: NCOR, 2009), 3–7, http://icbo.buffalo.edu/2009/Proceedings.pdf, accessed December 18, 2014.

5. See http://www.ebi.ac.uk/chebi/, accessed June 16, 2011.

6. Cornelius Rosse, Anand Kumar, Jose Leonardo V. Mejino, Daniel L. Cook, Landon T. Detwiler, and Barry Smith, "A Strategy for Improving and Integrating Biomedical Ontologies," in *Proceedings of AMIA Symposium* (Washington, DC: AMIA, 2005), 639–643.

7. We assume for the sake of simplicity that each ontology has a single root node. The Gene Ontology has three distinct roots, but is for this reason best conceived as a collection of three ontologies; see One Ontology . . . or Three? under "Ontology Structure," n.d., http://geneontology.org/page/ontology-structure, accessed August 5, 2014.

8. Aldo Gangemi, Nicola Guarino, Claudio Masolo, Alessandro Oltramari, Luc Schneider, "Sweetening Ontologies with DOLCE," in *Knowledge Engineering and Knowledge Management: Ontologies and the Semantic Web*, ed. Nicola Guarino (Berlin: Springer-Verlag, 2002), 166–181.

9. Ian Niles and Adam Pease, "Towards a Standard Upper Ontology," in *Proceedings of the International Conference on Formal Ontology in Information Systems* (FOIS) (New York: ACM Digital Press, 2002), 2–9.

3 Principles of Best Practice I

1. Foundational Model of Anatomy (FMA), http://sig.biostr.washington.edu/projects/fm/AboutFM.html, accessed September 1, 2014.

2. Mariana Casella dos Santos, James Matthew Fielding, Christoffel Dhaen, and Werner Ceusters, "Philosophical Scrutiny for Run-Time Support of Application Ontology Development," in *Proceedings of the International Conference on Formal Ontology and Information Systems* (FOIS), ed. Achille C. Varzi and Laure Vieu (Amsterdam: IOS Press, 2004), 342–352.

4 Principles of Best Practice II

1. Medical Subject Headings (MeSH), curated by the National Library of Medicine, http://www.ncbi.nlm.nih.gov/mesh, accessed September 1, 2014.

2. There may be occasional exceptions to this rule. For example, in chemistry, terms such as "zeatins," "cations," "esters," and "nitrates" are used in addition to their respective singular forms to denote special families of molecule types.

3. The Protein Ontology (PRO). Screenshot generated June 23, 2011, http://pir.georgetown.edu/cgi-bin/pro/browser_pro?quick_browse=Methylated_forms.

4. National Cancer Institute (NCI) Thesaurus, http://www.nlm.nih.gov/research/umls/, accessed via UMLS Knowledge Source Server Version 2006AC, September 28, 2006.

5. Ibid.

6. See http://ncit.nci.nih.gov/ncitbrowser.

7. Nicola Guarino, "Avoiding IS-A Overloading: The Role of Identity Conditions in Ontology Design," in *International Conference on Spatial Information Theory: Cognitive and Computational Foundations of Geographic Information Science, Proceedings*, ed. Nicola Guarino (London: Elsevier, 1999), 221–234. See also Nicola Guarino, "Some Ontological Principles for Designing Upper Level Lexical Resources," in *Proceedings of the First International Conference on Language Resources and Evaluation*, ed. Nicola Guarino (London: Elsevier, 1998), 527–534.

8. Barry Smith, Jakob Köhler, and Anand Kumar, "On the Application of Formal Principles to Life Science Data: A Case Study in the Gene Ontology," in *Proceedings of Data Integration in the Life Sciences* (DILS 2004), ed. Erhard Rahm (Dordrecht: Springer, 2004), 79–94.

9. Foundational Model of Anatomy (FMA), http://sig.biostr.washington.edu/projects/fm/AboutFM.html, accessed September 1, 2014.

10. SNOMED 17.0.0, http://www.snomedbrowser.com/Codes/Details/88162007, accessed September 1, 2014.

11. SNOMED asserts further that *spiritual or religious belief* is a subclass of *religion/philosophy*, and that *religion/philosophy* is a subclass of *social context*. Each of these steps provides further examples

of ontological error. Compare Ludger Jansen, "Four Rules for Classifying Social Entities," in *Philosophy, Computing and Information Science*, ed. Ruth Hagengruber and Uwe Riss (London: Pickering & Chatto, 2014), 189–200.

12. Franz Baader, Ian Horrocks, and Ulrike Sattler, "Description Logics," in *Handbook of Knowledge Representation*, ed. Frank van Harmelen, Vladimir Lifschitz, and Bruce Porter (Amsterdam: Elsevier, 2007), 135–180.

13. To see how ontologies might be complemented by a system for keeping track of particulars, see Werner Ceusters and Barry Smith, "Strategies for Referent Tracking in Electronic Health Records," *Journal of Biomedical Informatics* 39, no. 3 (June 2006): 362–378.

14. All definitions taken from the Foundational Model of Anatomy (FMA), http://bioportal .bioontology.org, accessed September 1, 2014.

15. See Edward Swiderski, "Some Salient Features of Ingarden's Ontology," *Journal of the British Society for Phenomenology* 6, no. 2 (May 1975): 81–90.

16. HL7 Glossary, various contributors (eds.), HL7 Publishing Technical Committee. Published November 22, 2005, 8:05 pm. HL7 Version 3 Standard. HL7's practices in this respect have not improved, as we document in Barry Smith, Lowell Vizenor, and Werner Ceusters, "Human Action in the Healthcare Domain: A Critical Analysis of HL7's Reference Information Model," in *Johanssonian Investigations: Essays in Honour of Ingvar Johansson on His Seventieth Birthday*, ed. Christer Svennerlind, Jan Almäng, and Rögnvaldur Ingthorsson (Berlin/New York: de Gruyter, 2013), 554–573.

17. Ibid.

18. See David M. Armstrong, *Universals and Scientific Realism* (Cambridge: Cambridge University Press, 1978). A sparse theory of universals is one that denies that there is a universal corresponding to every meaningful predicate. See also Barry Smith, "Against Fantology," in *Experience and Analysis*, ed. M. Reicher and J. Marek (Vienna: Hölder-Pichler-Tempsky, 2005), 153–170.

19. Werner Ceusters, Peter Elkin, and Barry Smith, "Negative Findings in Electronic Health Records and Biomedical Ontologies: A Realist Approach," *International Journal of Medical Informatics* 76 (2007): S326–S333.

20. For more on issues surrounding negation and negative vs. positive definitions, see Barry Smith and Werner Ceusters, "Ontological Realism: A Methodology for Coordinated Evolution of Scientific Ontologies," *Applied Ontology* 5, no. 3–4 (2010): 139–188.

21. SNOMED-CT, http://www.snomedbrowser.com/Codes/Details/8392000, accessed September 6, 2014.

22. Gramene Plant Environment Ontology, http://archive.gramene.org/db/ontology/search?id =EO:0007359, accessed September 6, 2014.

23. See, for example, Marianne Shaw, Todd Detwiler, Jim Brinkley, and Dan Suciu, "Generating Application Ontologies from Reference Ontologies," *Proceedings, American Medical Informatics Association Fall Symposium* (2008): 672–676.

24. See Alan L. Rector, "Modularisation of Domain Ontologies Implemented in Description Logics and Related Formalisms Including Owl," in *K-CAP '03: Proceedings of the 2nd International Conference on Knowledge Capture* (New York: ACM, 2003), 121–128.

5 Introduction to Basic Formal Ontology I

The material in this chapter and in chapters 6 and 7 is intended as an introduction to the categories and relations in BFO version 2.0 based on the documentation at http://ifomis.org/bfo/. This includes formal definitions of the terms introduced in these chapters as well as associated axioms and theorems and further explanatory material.

1. A more formal and rigorous treatment of the notions of "maximal" and "causal unity" is provided in the BFO 2.0 Specification referred to above.

2. Peter Simons talks in this connection of the parts inside an object being "welded" together in "Real Wholes, Real Parts: Mereology without Algebra," *Journal of Philosophy* 103, no. 5 (2006): 597–613. The case of blood cells shows, however, that not all parts of an object need be connected to the remainder.

3. Stefan Schulz, Anand Kumar, and Thomas Bittner, "Biomedical Ontologies: What *part-of* Is and Isn't," *Journal of Biomedical Informatics* 39, no. 3 (2006): 350–361.

4. Discussed in the following papers by Lars Vogt, Peter Grobe, Björn Quast, and Thomas Bartolomaeus: "Top-Level Categories of Constitutively Organized Material Entities—Suggestions for a Formal Top-Level Ontology," *PLoS ONE* 6, no. 4 (2011): e18794; "Accommodating Ontologies to Biological Reality—Top-Level Categories of Cumulative-Constitutively Organized Material Entities," *PLoS ONE* 7, no. 1 (2012): e30004; and " Fiat or Bona Fide Boundary—A Matter of Granular Perspective," *PLoS ONE* 7, no. 12 (2012): e48603.

5. Kevin Mulligan, "Relations—Through Thick and Thin," *Erkenntnis* 48 (1998): 325–353.

6. See Ludger Jansen, "The Ontology of Tendencies and Medical Information Science," *The Monist* 90 (2007): 534–555.

7. A proposal according to which *function* would be recognized by BFO as a sibling of *realizable entity* rather than of *disposition* is currently under review. See Johannes Röhl and Ludger Jansen, "Why Functions Are Not Special Dispositions: An Improved Classification of Realizables in Top-Level Ontologies," *Journal of Biomedical Semantics* 5, no. 27 (2014), 77–89.

8. There is a large and complex philosophical literature on functions; a useful overview for our purposes here is Jerome C. Wakefield, "Biological Function and Dysfunction," *Handbook of Evolutionary Psychology*, ed. David M. Buss (New York: Wiley, 2005), 878–902.

9. Röhl and Jansen, "Why Functions Are Not Special Dispositions."

10. See http://www.obofoundry.org/cgi-bin/detail.cgi?id=OGMS, accessed September 14, 2014, and Richard H. Scheuermann, Werner Ceusters, and Barry Smith, "Toward an Ontological Treatment of Disease and Diagnosis," in *Proceedings of the 2009 AMIA Summit on Translational Bioinformatics* (Washington, DC: AMIA, 2009), 116–120.

11. Consider also the case of complementary dispositions in the realm of infectious diseases discussed in Albert Goldfain, Barry Smith, and Lindsay G. Cowell, "Dispositions and the Infectious Disease Ontology," in *Formal Ontology in Information Systems: Proceedings of the Sixth International Conference* (FOIS 2010), ed. Antony Galton and Riichiro Mizoguchi (Amsterdam: IOS Press, 2010), 400–413. As OGMS is developed by extension from BFO, so IDO (the Infectious Disease Ontology) is developed in its turn from OGMS.

12. See http://www.ontobee.org/browser/index.php?o=IAO, last accessed September 29, 2014.

13. See http://www.sequenceontology.org/, last accessed September 29, 2014.

14. For a detailed consideration of the nature of literary works and the ways in which they are concretized, see Roman Ingarden, *The Literary Work of Art* (Evanston, IL: Northwestern University Press, 1974).

15. Cornelius Rosse and J. L. V. Mejino Jr., "The Foundational Model of Anatomy Ontology," in *Anatomy Ontologies for Bioinformatics: Principles and Practice*, vol. 6, ed. A. Burger, D. Davidson, and R. Baldock (London: Springer, 2007), 59–117.

16. On this topic see Avrum Stroll, *Surfaces* (Minneapolis: University of Minnesota Press, 1988); and Peter Simons, "Faces, Boundaries, and Thin Layers," in *Certainty and Surface in Epistemology and Philosophical Method*, Problems in Contemporary Philosophy, vol. 32 (Lewiston, NY: Mellen Press, 1991).

17. Arthur Eddington, *The Nature of the Physical World* (Cambridge: 1928), viii.

18. See, for example, Anand Kumar, Barry Smith, and Daniel Novotny, "Biomedical Informatics and Granularity," *Functional and Comparative Genomics* 5 (2004): 501–508.

19. Jonathan P. Bona, Jenny Rouleau, and Alan Ruttenberg. "Representing Modification Sites in PRO," *Proceedings of the 2014 International Conference on Biomedical Ontology* (CEUR Proceedings, vol. 1327), 2015, http://ceur-ws.org/Vol-1327/icbo2014_paper_46.pdf, accessed May 6, 2015.

20. See, for example, Brandon Bennett, "Space, Time, Matter and Things," in *Formal Ontology in Information Systems: Proceedings of the Fourth International Conference* (FOIS 2001), ed. C. Welty and B. Smith (New York: ACM, 2001), 105–116.

21. See Anthony Cohn and Achille Varzi, "Mereotopological Connection," *Journal of Philosophical Logic* 32, no. 4 (2003): 357–390.

22. See Barry Smith, "Classifying Processes: An Essay in Applied Ontology," *Ratio* 25, no. 4 (2012): 463–488.

6   Introduction to Basic Formal Ontology II

1. On this, see Barry Smith, "Classifying Processes: An Essay in Applied Ontology," *Ratio* 25, no. 4 (2012): 463–488.

2. See Selja Seppälä, Barry Smith, and Werner Ceusters, "Applying the Realism-Based Ontology Versioning Method for Tracking Changes in the Basic Formal Ontology," in *Formal Ontology in Information Systems: Proceedings of FOIS 2014* (Amsterdam: IOS Press, 2014), 227–240.

3. See http://obofoundry.org/, accessed December 17, 2014.

7   The Ontology of Relations

1. See also the OBO Relations Ontology Usage Examples Page, http://obofoundry.org/ro/, accessed September 14, 2014.

2. See, for example, Michael Ghiselin, *Metaphysics and the Origin of Species* (Albany: State University of New York Press, 1997), and David L. Hull, "Are Species Really Individuals?," *Systematic Zoology* 25 (1976): 174–191, and n. 25 (chapter 1, this volume).

3. Smith et al., "Relations in Biomedical Ontologies," *Genome Biology* 6, no. 5 (2005), accessed September 25, 2014, doi:10.1186/gb-2005-6-5-r46.

4. See http://www.obofoundry.org/ro/, accessed December 14, 2014.

5. See D. A. Randell, Z. Cui, and A. G. Cohn, "A Spatial Logic Based on Regions and Connection," in *Proceedings of the 3rd International Conference on Knowledge Representation and Reasoning*, ed. D. A. Randell, Z. Cui, and A. G. Cohn (San Mateo: Morgan Kaufmann, 1992), 165–176. See also Anthony G. Cohn, Brandon Bennett, John Gooday, and Nicholas Mark Gotts, "Qualitative Spatial Representation and Reasoning with the Region Connection Calculus," *GeoInformatica* 1 (1997): 275–316.

6. Robert Battle and Dave Kolas, "Enabling the Geospatial Semantic Web with Parliament and GeoSPARQL," *Semantic Web* 3, no. 4 (2012): 355–370.

7. James F. Allen, "Maintaining Knowledge about Temporal Intervals," *Communications of the ACM* 26, no. 11 (1983): 832–843.

8. Cornelius Rosse and José L. V. Mejino Jr., "The Foundational Model of Anatomy Ontology," in *Anatomy Ontologies for Bioinformatics: Principles and Practice*, vol. 6, ed. A. Burger, D. Davidson, and R. Baldock (London: Springer, 2007, 59–117. Christine Golbreich, Songmao Zhang, and Olivier Bodenreider, "The Foundational Model of Anatomy in OWL 2 and Its Use," *Artificial Intelligence in Medicine* 57, no. 2 (2013): 119–132.

9. Cornelius Rosse and Jose L. V. Mejino Jr., "A Reference Ontology for Biomedical Informatics: the Foundational Model of Anatomy," *Journal of Biomedical Informatics* 36 (2003): 478–500.

10. See http://www.obo foundry.org/ro/, accessed December 14, 2014.

11. A first-order logic release of BFO version 2 is in preparation. For an introduction to first-order logic, see Wilfred Hodges, "Classical Logic I: First Order Logic," in *The Blackwell Guide to Philosophical Logic*, ed. Lou Goble (Oxford: Blackwell, 2001), 9–32.

8   Basic Formal Ontology at Work

1. See http://protege.stanford.edu/, accessed December 14, 2014.

2. For an overview, see the OWL 2 Web Ontology Language Document Overview (second edition), http://www.w3.org/TR/owl2-overview/, accessed September 27, 2014.

3. Tim Berners-Lee, James Hendler, and Ora Lassila, "The Semantic Web," *Scientific American* (May 2001).

4. Protégé has a useful tutorial titled "Protégé OWL Tutorial," written by Matthew Horridge and others, http://130.88.198.11/tutorials/protegeowltutorial/, accessed December 14, 2014.

5. See "OWL Web Ontology Language Overview: W3C Recommendation," http://www.w3.org/TR/owl-features/; also see W3C's "DAML+OIL (March 2001) Reference Description," http://www.w3.org/TR/daml+oil-reference, accessed December 14, 2014.

6. See the W3C website, http://www.w3.org, accessed December 14, 2014.

7. "OWL Web Ontology Language Overview: W3C Proposed Recommendation," http://www.w3.org/TR/2003/PR-owl-features-20031215/. See also the press release at http://www.w3.org/2004/01/sws-pressrelease, accessed December 14, 2014.

8. See the W3C's "A History of HTML," http://www.w3.org/People/Raggett/book4/ch02.html, accessed December 10, 2014.

9. http://www.w3.org/MarkUp/; also see the W3C's "XML: Development History," http://www.w3.org/XML/hist2002, accessed December 10, 2014.

10. See the W3C's "Resource Description Framework (RDF) Model and Syntax Specification," http://www.w3.org/TR/1999/REC-RDFSyntax-19990222/, accessed December 10, 2014.

11. Ibid.

12. See Bernardo Grau et al., "OWL 2: The Next Step for OWL," *Web Semantics: Science, Services and Agents on the World Wide Web* 6, no. 4 (2008): 309–322, http://www.sciencedirect.com/science/article/pii/S1570826808000413, accessed December 10, 2014.

13. The problems with RDF and RDFS just mentioned, as well as others, are discussed in the W3C document "Web Ontology Language (OWL) Use Cases and Requirements," http://www.w3.org/TR/2003/WD-webont-req-20030331/, accessed December 14, 2014; also see Grigoris Antoniou and Frank van Harmelen, "Web Ontology Language: OWL," in *Handbook on Ontologies*, ed. Steffan Staab and Rudi Studer (Berlin: Springer, 2009), 91–110.

14. http://www.w3.org/TR/sparql11-overview/, accessed December 14, 2014.

15. See the W3C's document "OWL Web Ontology Language: Test Cases," http://www.w3.org/TR/2004/REC-owl-test-20040210/, accessed December 14, 2014.

16. See Grau et al., "OWL 2."

17. Taken from http://www.ifomis.org/bfo/users, accessed December 14, 2014.

18. http://www.obofoundry.org/cgi-bin/detail.cgi?id=OGMS, accessed December 14, 2014.

19. See http://www.obofoundry.org/cgi-bin/detail.cgi?id=OGMS, accessed December 14, 2014.

20. http://infectiousdiseaseontology.org/page/Main_Page, accessed December 14, 2014.

21. http://www.ontobee.org/browser/index.php?o=IAO, accessed December 14, 2014.

22. http://www.ontobee.org/browser/index.php?o=MF, accessed December 14, 2014.

23. http://www.ontobee.org/browser/index.php?o=MFOEM, accessed December 14, 2014.

24. http://obofoundry.org/, accessed December 14, 2014.

Appendix

1. http://www.ontobee.org/, accessed September 1, 2014.

2. See Melanie Courtot et al., "MIREOT: The Minimum Information to Reference an External Ontology Term," *Applied Ontology* 6, no. 1 (January 2011): 23–33.

Bibliography

Adams, E. "Topology, Empiricism, and Operationalism." *Monist* 79 (1996): 1–20.

Allen, James F. "Maintaining Knowledge about Temporal Intervals." *Communications of the ACM* 26 (11) (November 1983): 832–843.

Antoniou, Grigoris, and Frank van Harmelen. "Web Ontology Language: OWL." In *Handbook on Ontologies*, ed. Steffan Staab and Rudi Studer, 91–110. Berlin: Springer, 2009.

Ariew, A., R. Cummins, and M. Perlman, eds. *Functions: New Essays in the Philosophy of Biology and Psychology*. Oxford: Oxford University Press, 2002.

Aristotle. *The Complete Works of Aristotle*. Ed. Jonathan Barnes. Princeton: Princeton University Press, 1995.

Armstrong, David. *Universals and Scientific Realism*. Cambridge: Cambridge University Press, 1978.

Armstrong, David. *Universals: An Opinionated Introduction*. Boulder, CO: Westview Press, 1989.

Arp, Robert. "Realism and Antirealism in Informatics Ontologies." *The American Philosophical Association: Philosophy and Computers* 9 (1) (2009): 19–23.

Arp, Robert. "Philosophical Ontology, Domain Ontology, and Formal Ontology." In *Key Terms in Logic*, ed. Jon Williamson and Federica Russo, 74–75. London: Continuum, 2010.

Arp, Robert, Rethy Chhem, Cesare Romagnoli, and James Overton. "Radiological and Biomedical Knowledge Integration: The Ontological Way." In *Radiology Education: The Scholarship of Teaching and Learning*, ed. Rethy Chhem, Kathy Hilbert, and Teresa Van Deven, 87–104. Berlin: Springer, 2009.

Baader, Franz, Diego Calvanese, Deborah L. McGuinness, Daniele Nardi, and Peter F. Patel-Schneider. *The Description Logic Handbook: Theory, Implementation and Applications*. Cambridge: Cambridge University Press, 2010.

Baader, Franz, Ian Horrocks, and Ulrike Sattler. "Description Logics." In *Handbook of Knowledge Representation*, ed. Frank van Harmelen, Vladimir Lifschitz, and Bruce Porter, 135–180. Amsterdam: Elsevier, 2007.

Barnes, Jonathan, ed. *Porphyry: Introduction*. Oxford: Oxford University Press, 2006.

Batchelor, Colin, Janna Hastings, and Christoph Steinbeck. "Ontological Dependence, Dispositions and Institutional Reality in Chemistry." In *Formal Ontology in Information Systems: Proceedings of the Sixth International Conference (FOIS 2010)*, ed. Antony Galton and Riichiro Mizoguchi, 271–284. Amsterdam: IOS Press, 2010.

Battle, Robert, and Dave Kolas. "Enabling the Geospatial Semantic Web with Parliament and GeoSPARQL." *Semantic Web* 3 (4) (2012): 355–370.

Bennett, Brandon. "Space, Time, Matter and Things." In *Formal Ontology in Information Systems: Proceedings of the Fourth International Conference (FOIS 2001)*, ed. C. Welty and B. Smith, 105–116. New York: ACM, 2001.

Bennett, Brandon, V. Chaudhri, and N. Dinesh. "A Vocabulary of Topological and Containment Relations for a Practical Biological Ontology." In *Spatial Information Theory: Proceedings of COSIT 2013*, Lecture Notes in Computer Science, vol. 8116, ed. J. Stell, T. Tenbrink and Z. Wood, 418–437. Scarborough, UK: Springer, 2014.

Berners-Lee, Tim, James Hendler, and Ora Lassila. "The Semantic Web." *Scientific American* (May) (2001): 17.

Bird, A. *Nature's Metaphysics: Laws and Properties*. Oxford: Oxford University Press, 2007.

Bittner, Thomas. "A Mereological Theory of Frames of Reference." *International Journal of Artificial Intelligence Tools* 13 (1) (2004): 171–198.

Bittner, Thomas, and Maureen Donnelly. "Logical Properties of Foundational Relations in Bio-ontologies." *Artificial Intelligence in Medicine* 39 (2007): 197–216.

Bittner, Thomas, and Maureen Donnelly. "A Temporal Mereology for Distinguishing between Integral Objects and Portions of Stuff." In *Proceedings of the Twenty-Second AAAI Conference on Artificial Intelligence (AAAI)*, ed. R. Holte and A. Howe, 287–292. London, Elsevier: 2007.

Bittner, Thomas, and Barry Smith. "A Theory of Granular Partitions." In *Applied Ontology: An Introduction*, ed. Katherine Munn and Barry Smith, 125–158. Frankfurt: Ontos Verlag, 2008.

Bittner, Thomas, Maureen Donnelly, and Barry Smith. "Individuals, Universals, Collections: On the Foundational Relations of Ontology." In *Formal Ontology and Information Systems: Proceedings of FOIS 2004*, ed. Achille Varzi and Laure Vieu, 37–48. Amsterdam: IOS Press, 2004.

Bodenreider, Olivier. "Circular Hierarchical Relationships in the UMLS: Etiology, Diagnosis, Treatment, Complications and Prevention." *Proceedings of the American Medical Informatics Association Symposium* 23 (2001): 57–61.

Bona, Jonathan P., Jenny Rouleau, and Alan Ruttenberg. "Representing Modification Sites in PRO." *Proceedings of the 2014 International Conference on Biomedical Ontology* (CEUR Proceedings, vol. 1327). 2015. Accessed May 6, 2015. http://ceur-ws.org/Vol-1327/icbo2014_paper_46.pdf.

Borges, Jorge Luis. "The Analytical Language of John Wilkins." In *Other Inquisitions: 1937–1952*. Austin: University of Texas Press, 2000.

Bourget, David, and David Chalmers, eds. "The PhilPapers Survey." *PhilPapers*. n.d. Accessed August 15, 2014. http://philpapers.org/surveys/.

Brachman, Ronald J., and Hector J. Levesque, eds. *Readings in Knowledge Representation*. San Francisco: Morgan Kaufmann Publishers Inc., 1985.

Casati, Roberto, Barry Smith, and Achille Varzi. "Ontological Tools for Geographic Representation." In *Formal Ontology in Information Systems: Proceedings of the First International Conference (FOIS 1998)*, ed. Nicola Guarino, 77–85. Amsterdam: IOS Press, 1998.

Casati, Roberto, and Achille Varzi eds. *Events*. Dartmouth: Aldershot, 1996.

Casati, Roberto, and Achille Varzi. *Holes and Other Superficialities*. Cambridge, MA: MIT Press, 1994.

Casati, Roberto, and Achille Varzi. *Parts and Places: The Structures of Spatial Representation*. New York: Bradford Books, 1999.

Casella dos Santos, Mariana, James Matthew Fielding, Christoffel Dhaen, and Werner Ceusters. "Philosophical Scrutiny for Run-Time Support of Application Ontology Development." In *Proceedings of the International Conference on Formal Ontology and Information Systems (FOIS)*, ed. Achille C. Varzi and Laure Vieu, 342–352. Amsterdam: IOS Press, 2004.

Ceusters, Werner, Peter Elkin, and Barry Smith. "Negative Findings in Electronic Health Records and Biomedical Ontologies: A Realist Approach." *International Journal of Medical Informatics* 76 (2007): S326–S333.

Ceusters, Werner, and Barry Smith. "A Realism-Based Approach to the Evolution of Biomedical Ontologies." In *Proceedings of the AMIA Symposium*, 121–125. Washington, DC: AMIA, 2006.

Ceusters, Werner, Barry Smith, and Louis Goldberg. "A Terminological and Ontological Analysis of the NCI Thesaurus." *Methods of Information in Medicine* 44 (2005): 498–507.

Ceusters, Werner, Barry Smith, Anand Kumar, and C. Dhaen. "Mistakes in Medical Ontologies: Where Do They Come From and How Can They Be Detected?" *Studies in Health Technology and Informatics* 102 (2004): 145–164.

Ceusters, Werner, F. Steurs, P. Zanstra, E. Van Der Haring, and Jeremy Rogers. "From a Time Standard for Medical Informatics to a Controlled Language for Health." *International Journal of Medical Informatics* 48 (1–3) (1998): 85–101.

Chute, Christopher G. "Medical Concept Representation." In *Medical Informatics: Integrated Series in Information Systems*, vol. 8, ed. H. Chen, S. S. Fuller, C. Friedman, and W. Hersh, 163–182. New York: Springer, 2005.

Cimiano, Philipp, Christina Unger, and John McCrae. *Ontology-Based Interpretation of Natural Language*. San Rafael, CA: Morgan & Claypool, 2014.

Cimino, James J. "In Defense of the Desiderata." *Journal of Biomedical Informatics* 39 (3) (2006): 299–306.

Clarke, B. L. "A Calculus of Individuals Based on 'Connection.'" *Notre Dame Journal of Formal Logic* 23 (3) (July 1981): 204–218.

Cohn, Anthony G., Brandon Bennett, John Gooday, and Nicholas Mark Gotts. "Qualitative Spatial Representation and Reasoning with the Region Connection Calculus." *GeoInformatica* 1 (1997): 275–316.

Cohn, Anthony G., and J. Renz. "Qualitative Spatial Representation and Reasoning." In *Handbook of Knowledge Representation*, ed. F. van Harmelen, V. Lifschitz, and B. Porter, 551–596. Amsterdam: Elsevier, 2008.

Cohn, Anthony G., and Achille Varzi. "Mereotopological Connection." *Journal of Philosophical Logic* 32 (4) (2003): 357–390.

Courtot, Mélanie, Frank Gibson, Allyson L. Lister, et al. "MIREOT: The Minimum Information to Reference an External Ontology Term." *Applied Ontology* 6 (1) (January 2011): 23–33.

Dennett, Daniel. *Brainchildren: Essays on Designing Minds.* Cambridge, MA: MIT Press, 1998.

DeSalvo, Karen B., and Erica Galvez. "Connecting Health and Care for the Nation: A 10-Year Vision to Achieve an Interoperable Health IT Infrastructure." The Office of the National Coordinator for Health Information Technology. Last updated June 2014. Accessed September 1, 2014. http://www.healthit.gov/sites/default/files/ONC10year InteroperabilityConceptPaper.pdf.

Dipert, Randall. *Artifacts, Art Works, and Agency.* Philadelphia: Temple University Press, 1993.

Donnelly, Maureen. "Containment Relations in Anatomical Ontologies." In *Proceedings of the AMIA Symposium*, 206–210. London: Elsevier, 2005.

Donnelly, Maureen. "A Formal Theory for Reasoning about Parthood, Connection, and Location." *Artificial Intelligence* 160 (2004): 145–172.

Donnelly, Maureen. "Relative Places." In *Formal Ontology in Information Systems: Proceedings of the Fourth International Conference (FOIS 2004)*, ed. Achille Varzi and Laure Vieu, 249–260. Amsterdam: IOS Press, 2004.

Donnelly, Maureen, Thomas Bittner, and Cornelius Rosse. "A Formal Theory for Spatial Representation and Reasoning in Biomedical Ontologies." *Artificial Intelligence in Medicine* 36 (2006): 1–27.

dos Santos, Mariana, James Matthew Fielding, Christoffel Dhaen, and Werner Ceusters. "Philosophical Scrutiny for Run-Time Support of Application Ontology Development." In *Proceedings of the International Conference on Formal Ontology and Information Systems (FOIS)*, ed. Achille C. Varzi and Laure Vieu, 342–352. Amsterdam: IOS Press, 2004.

Dretske, Fred. "Can Events Move?" *Mind* 76 (1967): 479–492.

Eddington, Arthur. *The Nature of the Physical World*. Cambridge: Cambridge University Press, 1928.

Ereshefsky, Marc. "Species." *The Stanford Encyclopedia of Philosophy*. Spring 2010 edition, ed. Edward N. Zalta. Accessed August 5, 2014. http://plato.stanford.edu/archives/spr2010/entries/species/.

Feigenbaum, Lee, Ivan Herman, Tonya Hongsermeier, Eric Neumann, and Susie Stephens. "The Semantic Web in Action." *Scientific American* 297 (2007): 90–97.

Fine, Kit. "Ontological Dependence." *Proceedings of the Aristotelian Society, New Series* 95 (1995): 269–290.

First Healthcare Interoperability Resources (FHIR). Last updated September 30, 2014. Accessed February 2, 2015. http://hl7.org/implement/standards/fhir/overview.html.

FOAF Vocabulary Specification 0.99. Last updated January 2014. Accessed September 1, 2014. http://xmlns.com/foaf/spec/#term_Document.

Franklin, J. "Stove's Discovery of the Worst Argument in the World." *Philosophy* 77 (2002): 615–624.

Galton, Anthony. *Qualitative Spatial Change*. Oxford: Oxford University Press, 2001.

Galton, Anthony, and Riichiro Mizoguchi. "The Water Falls But the Waterfall Does Not Fall: New Perspectives on Objects, Processes, and Events." *Applied Ontology* 4 (2) (2009): 71–107.

Gangemi, Aldo, Nicola Guarino, Claudio Masolo, Alessandro Oltramari, and Luc Schneider. "Sweetening Ontologies with DOLCE." In *Knowledge Engineering and Knowledge Management: Ontologies and the Semantic Web*, vol. 2473, ed. Nicola Guarino, 166–181. Berlin: Springer-Verlag, 2002.

Geller, James. "What Is an Ontology?" n.d. Accessed August 4, 2014. http://web.njit.edu/~geller/what_is_an_ontology.html.

Ghiselin, Michael. *Metaphysics and the Origin of Species*. Albany: State University of New York Press, 1997.

Golbreich, Christine, Songmao Zhang, and Olivier Bodenreider. "The Foundational Model of Anatomy in OWL 2 and Its Use." *Artificial Intelligence in Medicine* 57 (2) (2013): 119–132.

Goldfain, Albert, Barry Smith, and Lindsay G. Cowell. "Constructing a Lattice of Infectious Disease Ontologies from a Staphylococcus aureus Isolate Repository." *Proceedings of the Third International Conference on Biomedical Ontology (CEUR 897)*, Graz, July 21–25, 2012. Accessed September 1, 2014. http://ceur-ws.org/Vol-897/.

Goldfain, Albert, Barry Smith, and Lindsay G. Cowell. "Dispositions and the Infectious Disease Ontology." In *Formal Ontology in Information Systems: Proceedings of the Sixth International Conference (FOIS 2010)*, ed. Antony Galton and Riichiro Mizoguchi, 400–413. Amsterdam: IOS Press, 2010.

Goldfain, Albert, Barry Smith, and Lindsay G. Cowell. "Towards an Ontological Representation of Resistance: The Case of MRSA." *Journal of Biomedical Informatics* 44 (1) (February 2011): 35–41.

Grau, Bernardo, Ian Horrocks, Boris Motik, Bijan Parsia, Peter Patel-Schneider, and Ulrike Sattler. "OWL 2: The Next Step for OWL." *Web Semantics: Science, Services, and Agents on the World Wide Web* 6 (4) (2008): 309–322.

Grenon, Pierre. "The Formal Ontology of Spatio-Temporal Reality and Its Formalization." In *Foundations and Applications of Spatio-Temporal Reasoning*, ed. H. Guesguen, D. Mitra, and J. Renz, 27–34. Amsterdam: AAAI Press, 2003.

Grenon, Pierre. "A Primer on Knowledge Management and Ontological Engineering." In *Applied Ontology: An Introduction*, ed. Katherine Munn and Barry Smith, 57–82. Frankfurt: Ontos Verlag, 2008.

Grenon, Pierre, and Barry Smith. "A Formal Theory of Substances, Qualities and Universals." In *Proceedings of the International Conference on Formal Ontology and Information Systems (FOIS 2004)*, ed. Achille Varzi and Laure Vieu, 49–59. Amsterdam: IOS Press, 2004.

Grenon, Pierre, and Barry Smith. "SNAP and SPAN: Towards Dynamic Spatial Ontology." *Spatial Cognition and Computation* 4 (1) (2004): 1–10.

Grenon, Pierre, Barry Smith, and Louis Goldberg. "Biodynamic Ontology: Applying BFO in the Biomedical Domain." In *Ontologies in Medicine*, ed. D. Pisanelli, 20–38. Amsterdam: IOS Press, 2004.

Grobe, Susan J. "ICNP Version 1: International Classification for Nursing Practice—A Unified Nursing Language System." 2005. Accessed August 30, 2014. www.nicecomputing.ch/nieurope/S%20Grobe%20ICNP.pdf.

Gruber, Tom. "A Translation Approach to Portable Ontologies." *Knowledge Acquisition* 5 (2) (1992): 199–220.

Gruber, Tom. "What Is an Ontology?" 1992. Accessed September 1, 2014. http://www-ksl.stanford.edu/kst/what-is-an-ontology.html.

Guarino, Nicola. "Avoiding IS-A Overloading: The Role of Identity Conditions in Ontology Design." In *International Conference on Spatial Information Theory: Cognitive and Computational Foundations of Geographic Information Science, Proceedings*, 221–234. London: Elsevier, 1999.

Guarino, Nicola. "Some Ontological Principles for Designing Upper Level Lexical Resources." In *Proceedings of the First International Conference on Language Resources and Evaluation*, ed. Nicola Guarino, 527–534. London: Elsevier, 1998.

Haemmerli, Marion, and Achille Varzi. "Adding Convexity to Mereotopology." In *Formal Ontology in Information Systems*, ed. Achille Varzi, 65–78. Amsterdam: IOS Press, 2014.

Hankinson, R. "Science." In *The Cambridge Companion to Aristotle*, ed. Jonathan Barnes, 140–167. Cambridge: Cambridge University Press, 1997.

Health Informatics. L7 version 3. Reference Information Model. Release 4. Document ISO/HL7 21731:2011(E). 2011. Accessed September 1, 2014. http://www.hl7.org/index.cfm.

Hennig, Boris. "Occurrents." In *Applied Ontology: An Introduction*, ed. Katherine Munn and Barry Smith, 255–284. Frankfurt: Ontos Verlag, 2008.

Hennig, Boris. "What Is Formal Ontology?" In *Applied Ontology: An Introduction*, ed. Katherine Munn and Barry Smith, 39–56. Frankfurt: Ontos Verlag, 2008.

Hill, David P., Barry Smith, Monica S. McAndrews-Hill, and Judith A. Blake. "Gene Ontology Annotations: What They Mean and Where They Come From." *BMC Bioinformatics* 9 (suppl. 5) (2008): S2.

Hitzler, Pascal, Markus Krötzsch, and Sebastian Rudolph. *Foundations of Semantic Web Technologies*. Boca Raton, FL: Chapman & Hall, 2009.

Hobbs, J. R., and R. C. Moore, eds. *Formal Theories of the Common-Sense World*. Norwood, NJ: Ablex, 1985.

Hodges, Wilfrid. "Classical Logic I: First Order Logic." In *The Blackwell Guide to Philosophical Logic*, ed. Lou Goble, 9–32. Oxford: Blackwell, 2001.

Horrocks, Ian. "Ontologies and the Semantic Web." *Communications of the ACM* 51 (12) (2008): 58–67.

Horrocks, Ian, Peter Patel-Schneider, Deborah McGuinness, and Christopher Welty. "Ontology Languages for the Semantic Web." In *The Description Logic Handbook*, ed. Franz Baader, Diego Calvanese, Deborah McGuinness, Daniele Nardi, and Peter Patel-Schneider, 458–486. Cambridge: Cambridge University Press, 2003.

Hull, David L. "Are Species Really Individuals?" *Systematic Zoology* 25 (1976): 174–191.

Ingarden, Roman. *The Literary Work of Art*. Evanston, IL: Northwestern University Press, 1974.

International Health Terminology Standards Development Organisation. *SNOMED CT® Technical Reference Guide—July 2010 International Release*. Washington, DC: College of American Pathologists, 2010.

ISO 1087-1:2000. Terminology Work—Vocabulary—Part 1: Theory and Application, 2000.

Jansen, Ludger. "Categories: The Top-Level Ontology." In *Applied Ontology: An Introduction*, ed. Katherine Munn and Barry Smith, 173–196. Frankfurt: Ontos Verlag, 2008.

Jansen, Ludger. "Classifications." In *Applied Ontology: An Introduction*, ed. Katherine Munn and Barry Smith, 159–172. Frankfurt: Ontos Verlag, 2008.

Jansen, Ludger. "Four Rules for Classifying Social Entities." In *Philosophy, Computing and Information Science*, ed. Ruth Hagengruber and Uwe Riss, 189–200. London: Pickering & Chatto, 2014.

Jansen, Ludger. "The Ontology of Tendencies and Medical Information Science." *The Monist* 90 (2007): 534–555.

Johansson, Ingvar. "Bioinformatics and Biological Reality." In *Applied Ontology: An Introduction*, ed. Katherine Munn and Barry Smith, 285–310. Frankfurt: Ontos Verlag, 2008.

Johansson, Ingvar. "Determinables as Universals." *Monist* 83 (2000): 101–121.

Johansson, Ingvar. *An Enquiry into the Categories of Nature, Man, and Society*. New York: Routledge, 1989.

Johansson, Ingvar. *Ontological Investigations: An Enquiry into the Categories of Nature, Man, and Society*. New York: Routledge, 1989.

Johansson, Ingvar, and Niels Lynøe. *Medicine and Philosophy: A Twenty-First Century Introduction*. Frankfurt: Ontos Verlag, 2009.

Koepsell, David, Robert Arp, Jennifer Fostel, and Barry Smith. "Creating a Controlled Vocabulary for the Ethics of Human Research: Towards a Biomedical Ethics Ontology." *Journal of Empirical Research on Human Research Ethics* 4 (2009): 43–58.

Köhler, Jacob, Katherine Munn, Alexander Ruegg, Andre Skusa, and Barry Smith. "Quality Control for Terms and Definitions in Ontologies and Taxonomies." *BMC Bioinformatics* 7 (2006): 212.

Kumar, Anand, and Barry Smith. "The Unified Medical Language System and the Gene Ontology: Some Critical Reflections." *KI 2003: Advances in Artificial Intelligence* 2821 (2003): 135–148.

Kumar, Anand, Barry Smith, and Daniel Novotny. "Biomedical Informatics and Granularity." *Functional and Comparative Genomics* 5 (2004): 501–508.

Low, H.-S., C. J. O. Baker, A. Garcia, and M. R. Wenk. "An OWL-DL Ontology for Classification of Lipids." In *Proceedings of the International Conference on Biomedical Ontology (ICBO 2009)*, 3–7. Buffalo, NY: NCOR, 2009. Accessed December 18, 2014. http://icbo.buffalo.edu/2009/Proceedings.pdf.

Lowe, E. J. *A Survey of Metaphysics*. Oxford: Oxford University Press, 2002.

Lowe, E. J. *The Four Category Ontology: A Metaphysical Foundation for Natural Science*. Oxford: Oxford University Press, 2006.

Martin, C. B. "Dispositions and Conditionals." *Philosophical Quarterly* 44 (1994): 1–8.

Masci, Anna M., Cecilia N. Arighi, Alexander D. Diehl, Anne E. Lieberman, Chris Mungall, Richard H. Scheuermann, Barry Smith, and Lindsay G. Cowell. "An Improved Ontological Representation of Dendritic Cells as a Paradigm for all Cell Types." *BMC Bioinformatics* 10 (70) (February 2009). doi:10.1186/1471-2105-10-70. Accessed September 29, 2014.

Mayr, E. "The Autonomy of Biology: The Place of Biology among the Sciences." *Quarterly Review of Biology* 71 (1996): 97–106.

"Microsoft HealthVault." Last updated 2014. Accessed August 4, 2014. http://msdn.microsoft.com/en-us/library/aa155110.aspx.

Miliard, Mik "Data Variety Bigger Hurdle than Volume." *HealthcareITNews*. July 3, 2014. Accessed August 25, 2014. http://www.healthcareitnews.com/news/data-variety-bigger-hurdle-volume ?topic=02,06&mkt_tok=3RkMMJWWfF9wsRonuq3IZKXonjHpfsX87OQkWbHr08Yy0EZ5VunJEU Wy2YlDT9Q%2FcOedCQkZHblFnVUKSK2vULcNqKwP.

Motik, Boris, Ian Horrocks, and Ulrike Sattler. "Bridging the Gap Between OWL and Relational Databases." *Journal of Web Semantics* 7 (2) (2009): 74–89.

Mulligan, Kevin. "Relations—Through Thick and Thin." *Erkenntnis* 48 (1998): 325–353.

Mulligan, Kevin, Peter M. Simons, and Barry Smith. "Truth-Makers." *Philosophy and Phenomenological Research* 44 (1984): 287–321.

Munn, Katherine. "Introduction: What Is Ontology For?" In *Applied Ontology: An Introduction*, ed. Katherine Munn and Barry Smith, 7–19. Frankfurt: Ontos Verlag, 2008.

Niles, Ian, and Adam Pease. "Towards a Standard Upper Ontology." In *Proceedings of the International Conference on Formal Ontology in Information Systems (FOIS)*, ed. Adam Pease. 2–9. New York: ACM Digital Press, 2002.

"Ontology Structure." n.d. Accessed August 5, 2014. http://www.geneontology.org/page/ontology-structure.

Randell, D. A., Z. Cui, and A. G. Cohn. "A Spatial Logic Based on Regions and Connection." In *Proceedings of the 3rd International Conference on Knowledge Representation and Reasoning*, 165–176. San Mateo, CA: Morgan Kaufmann, 1992.

Rector, Alan. "Modularisation of Domain Ontologies Implemented in Description Logics and Related Formalisms Including OWL." In *K-CAP '03: Proceedings of the 2nd International Conference on Knowledge Capture*, 121–128. New York: ACM, 2003.

Rector, Alan, Jeremy Roger, and Thomas Bittner. "Granularity, Scale and Collectivity: When Size Does and Does Not Matter." *Journal of Biomedical Informatics* 39 (2006): 333–349.

Robinson, Peter N., and Sebastian Bauer. *Introduction to Bio-ontologies*. New York: Chapman and Hall/CRC, 2011.

Rodriguez-Pereyra, G. *Resemblance Nominalism: A Solution to the Problem of Universals*. Oxford: Clarendon Press, 2002.

Röhl, Johannes, and Ludger Jansen. "Why Functions Are Not Special Dispositions: An Improved Classification of Realizables for Top-Level Ontologies." *Journal of Biomedical Semantics* 5 (27) (2014): 33–45

Rosenberg, A. *Darwinian Reductionism, or How to Stop Worrying and Love Molecular Biology*. Chicago: University of Chicago Press, 2006.

Rosse, Cornelius, Anand Kumar, Jose Leonardo V. Mejino, Daniel L. Cook, Landon T. Detwiler, and Barry Smith. "A Strategy for Improving and Integrating Biomedical Ontologies." In *Proceedings of the AMIA Symposium*, 639–643. Washington, DC: AMIA, 2005.

Rosse, Cornelius, and Jose L. V. Mejino Jr. "The Foundational Model of Anatomy Ontology." In *Anatomy Ontologies for Bioinformatics: Principles and Practice*, vol. 6, ed. Albert Burger, Duncan Davidson, and Richard Baldock, 59–117. Berlin: Springer, 2008.

Rosse, Cornelius, and Jose L. Mejino Jr. "A Reference Ontology for Biomedical Informatics: The Foundational Model of Anatomy." *Journal of Biomedical Informatics* 36 (2003): 478–500.

Scheuermann, Richard H., Werner Ceusters, and Barry Smith. "Toward an Ontological Treatment of Disease and Diagnosis." In *Proceedings of the 2009 AMIA Summit on Translational Bioinformatics*, 116–120. Washington, DC: AMIA, 2009.

Schulz, Stefan, Laszlo Balkanyi, Ronald Cornet, and Olivier Bodenreider. "From Concept Representations to Ontologies: A Paradigm Shift in Health Informatics?" *Healthcare Informatics Research* 19 (4) (2013): 235–242.

Schulz, Stefan, Anand Kumar, and Thomas Bittner. "Biomedical Ontologies: What *part-of* Is and Isn't." *Journal of Biomedical Informatics* 39 (3) (2006): 350–361.

Schwarz, Ulf, and Barry Smith. "Ontological Relations." In *Applied Ontology: An Introduction*, ed. Katherine Munn and Barry Smith, 219–234. Frankfurt: Ontos Verlag, 2008.

Searle, John. *The Construction of Social Reality*. New York: The Free Press, 1997.

Seppälä, Selja, Barry Smith, and Werner Ceusters. "Applying the Realism-Based Ontology Versioning Method for Tracking Changes in the Basic Formal Ontology." In *Formal Ontology in Information Systems: Proceedings of FOIS 2014*, 227–240. Amsterdam: IOS Press, 2014.

Shaw, Marianne, Landon T. Detwiler, James F. Brinkley, and Dan Suciu. "Generating Application Ontologies from Reference Ontologies." *Proceedings, American Medical Informatics Association Fall Symposium* (2008): 672–676.

Sider, Ted. *Four-Dimensionalism: An Ontology of Persistence and Time*. Oxford: Oxford University Press, 2005.

Silberstein, M., and J. McGeever. "The Search for Ontological Emergence." *Philosophical Quarterly* 49 (1999): 201–214.

Simons, Peter. "Particulars in Particular Clothing: Three Trope Theories of Substance." *Philosophy and Phenomenological Research* 54 (1994): 553–575.

Simons, Peter. "Continuants and Occurrents." *Proceedings of the Aristotelian Society* 74 (2000): 59–75.

Simons, Peter. "Faces, Boundaries, and Thin Layers." In *Certainty and Surface in Epistemology and Philosophical Method*, Problems in Contemporary Philosophy, vol. 32, 87–99. Lewiston, NY: Mellen Press, 1991.

Simons, Peter. *Parts: A Study in Ontology*. Oxford: Oxford University Press, 1997.

Simons, Peter. "Real Wholes, Real Parts: Mereology without Algebra." *Journal of Philosophy* 103 (5) (2006): 597–613.

Smith, Barry. "Against Fantology." In *Experience and Analysis*, ed. M. Reicher and J. Marek, 153–170. Vienna: Hölder-Pichler-Tempsky, 2005.

Smith, Barry. "Against Idiosyncrasy in Ontology Development." In *Formal Ontology and Information Systems: Proceedings of the Sixth International Conference (FOIS 2006)*, ed. B. Bennett and C. Fellbaum, 15–26. Amsterdam: IOS Press, 2006.

Smith, Barry. "The Benefits of Realism: A Realist Logic with Applications." In *Applied Ontology: An Introduction*, ed. Katherine Munn and Barry Smith, 109–124. Frankfurt: Ontos Verlag, 2008.

Smith, Barry. "Beyond Concepts: Ontology as Reality Representation." In *Formal Ontology in Information Systems: Proceedings of the Fourth International Conference (FOIS 2004)*, ed. Achille C. Varzi and Laure Vieu, 31–42. Amsterdam: IOS Press, 2004.

Smith, Barry. "Biometaphysics." In *Routledge Companion to Metaphysics*, ed. Robin Le Poidevin, Peter Simons, Andrew McGonigal, and Ross P. Cameron, 537–544. New York: Routledge, 2009.

Smith, Barry. "Boundaries: An Essay in Mereotopology." In *The Philosophy of Roderick Chisholm*, ed. Lewis Hahn, 534–561. LaSalle: Open Court, 1997.

Smith, Barry. "Classifying Processes: An Essay in Applied Ontology." *Ratio* 25 (4) (2012): 463–488.

Smith, Barry. "The Logic of Biological Classification and the Foundations of Biomedical Ontology." In *Invited Papers from the 10th International Conference in Logic Methodology and Philosophy of Science*, ed. Dag Westerståhl, 505–520. London: King's College Publications, 2005.

Smith, Barry. "Mereotopology: A Theory of Parts and Boundaries." *Data & Knowledge Engineering* 20 (1996): 287–303.

Smith, Barry. "New Desiderata for Biomedical Ontologies." In *Applied Ontology: An Introduction*, ed. Katherine Munn and Barry Smith, 84–107. Frankfurt: Ontos Verlag, 2008.

Smith, Barry. "On Classifying Material Entities in Basic Formal Ontology." In *Interdisciplinary Ontology (Proceedings of the Third Interdisciplinary Ontology Meeting)*, ed. Barry Smith, Riichiro Mizoguchi, and Sumio Nakagawa, 1–13. Tokyo: Keio University Press, 2012.

Smith, Barry. "On Substances, Accidents and Universals: In Defence of a Constituent Ontology." *Philosophical Papers* 26 (1997): 105–127.

Smith, Barry. "Ontology." In *Blackwell Guide to the Philosophy of Computing and Information*, ed. Luciano Floridi, 155–166. Oxford: Blackwell, 2003.

Smith, Barry. "Ontology (Science)." In *Ontology in Information Systems, Proceedings of the Fifth International Conference* (FOIS 2008), ed. C. Eschenbach and M. Gruninger, 21–35. Amsterdam: IOS Press, 2008.

Smith, Barry. Fiat Objects. *Topoi* 20 (2001): 131–148.

Smith, Barry, and Achille Varzi. "The Niche." *Noûs* 33 (2) (1999): 198–222.

Smith, Barry, and Achille Varzi. "Fiat and Bona Fide Boundaries." *Philosophy and Phenomenological Research* 60 (2000): 401–420.

Smith, Barry, and Achille Varzi. "Surrounding Space: The Ontology of Organism-Environment Relations." *Theory in Biosciences* 121 (2002): 139–162.

Smith, Barry, Michael Ashburner, Cornelius Rosse, Jonathan Bard, William Bug, Werner Ceusters, Louis J. Goldberg, Karen Eilbeck, Amelia Ireland, and Christopher J. Mungall, The OBI Consortium, Neocles Leontis, Philippe Rocca-Serra, Alan Ruttenberg, Susanna-Assunta Sansone, Richard H Scheuermann, Nigam Shah, Patricia L. Whetzel, and Suzanna Lewis. "The OBO Foundry: Coordinated Evolution of Ontologies to Support Biomedical Data Integration." *Nature Biotechnology* 25 (11) (November 2007): 1251–1255.

Smith, Barry, and Berit Brogaard. "A Unified Theory of Truth and Reference." *Logique et Analyse* 43 (169–170) (2003): 49–93.

Smith, Barry, and Werner Ceusters. "Ontological Realism: A Methodology for Coordinated Evolution of Scientific Ontologies." *Applied Ontology* 5 (3–4) (2010): 139–188.

Smith, Barry, and Werner Ceusters. "Strategies for Referent Tracking in Electronic Health Records." *Journal of Biomedical Informatics* 39 (3) (June 2006): 362–378.

Smith, Barry, Werner Ceusters, and Rita Temmerman. "Wüsteria." *Studies in Health Technology and Information* 116 (2005): 647–652.

Smith, Barry, and Werner Ceusters. "Towards Industrial Strength Philosophy: How Analytical Ontology Can Help Medical Informatics." *Interdisciplinary Science Reviews* 28 (2003): 106–111.

Smith, Barry, Werner Ceusters, Bert Klagges, Jacob Köhler, Anand Kumar, Jane Lomax, Chris Mungall, Fabian Neuhaus, Alan L. Rector, and Cornelius Rosse. "Relations in Biomedical Ontologies." *Genome Biology* 6 (5) (2005). doi:10.1186/gb-2005-6-5-r46. Accessed September 25, 2014.

Smith, Barry, Werner Ceusters, and Rita Temmerman. "Wüsteria." *Studies in Health Technology and Informatics* 116 (2005): 647–652.

Smith, Barry, and Pierre Grenon. "The Cornucopia of Formal Ontological Relations." *Dialectica* 58 (2004): 279–296.

Smith, Barry, and Bert Klagges. "Bioinformatics and Philosophy." In *Applied Ontology: An Introduction*, ed. Katherine Munn and Barry Smith, 21–38. Frankfurt: Ontos Verlag, 2008.

Smith, Barry, Jacob Köhler, and Anand Kumar. "On the Application of Formal Principles to Life Science Data: A Case Study in the Gene Ontology." In *Proceedings of Data Integration in the Life Sciences (DILS 2004)*, ed. Erhard Rahm, 79–94. Dordrecht: Springer, 2004.

Smith, Barry, and Anand Kumar. "On Controlled Vocabularies in Bioinformatics: A Case Study in the Gene Ontology." *BIOSILICO: Drug Discovery Today* 2 (2004): 246–252.

Smith, Barry, and Anand Kumar. "On Controlled Vocabularies in Bioinformatics: A Case Study in the Gene Ontology." *BIOSILICO: Drug Discovery Today* 2 (2004): 246–252.

Smith, Barry, Anand Kumar, and Thomas Bittner. "Basic Formal Ontology for Bioinformatics." IFOMIS Reports, 2005. Accessed December 14, 2014. http://philpapers.org/rec/KUMIR.

Smith, Barry, Waclaw Kusnierczyk, Daniel Schober, and Werner Ceusters. "Towards a Reference Terminology for Ontology Research and Development in the Biomedical Domain." In *Proceedings of the 2nd International Workshop on Formal Biomedical Knowledge Representation* (KR-MED 2006), vol. 222, ed. Olivier Bodenreider, 57–66. Baltimore, MD: KR-MED Publications, 2006. Accessed December 17, 2014. http://www.informatik.uni-trier.de/~ley/db/conf/krmed/krmed2006.html.

Smith, Barry, and Achille Varzi. "Surrounding Space: The Ontology of Organism-Environment Relations." *Theory in Biosciences* 121 (2002): 139–162.

Smith, Barry, Lowell Vizenor, and Werner Ceusters. "Human Action in the Healthcare Domain: A Critical Analysis of HL7's Reference Information Model." In *Johanssonian Investigations: Essays in Honour of Ingvar Johansson on His Seventieth Birthday*, ed. Christer Svennerlind, Jan Almäng, and Rögnvaldur Ingthorsson, 554–573. Berlin/New York: de Gruyter, 2013.

Stroll, Avrum. *Surfaces*. Minneapolis: University of Minnesota Press, 1988.

Swiderski, Edward. "Some Salient Features of Ingarden's Ontology." *Journal of the British Society for Phenomenology* 6 (2) (May 1975): 81–90.

U.S. Army. "Joint Doctrine Hierarchy." n.d. Accessed August 5, 2014. http://usacac.army.mil/cac2/doctrine/CDM%20pages/cdm_joint%20heirarchy.html.

U.S. Department of Health and Human Services. "Development of Software and Analysis Methods for Biomedical Big Data in Targeted Areas of High Need (U01)." 2014. Accessed August 25, 2014. http://grants.nih.gov/grants/guide/rfa-files/RFA-HG-14-020.html.

van Harmelen, Frank. "Web Ontology Language: OWL." In *Handbook on Ontologies*, ed. Steffan Staab and Rudi Studer, 91–110. Berlin: Springer, 2009.

Van Heijst, G., A. T. Schreiber, and B. J. Wielinga. "Using Explicit Ontologies in KBS Development." *International Journal of Human–Computer Studies* 45 (1996): 183–192.

Varzi, Achille. "Boundaries, Continuity, and Contact." *Noûs* 31 (1997): 26–58.

Vogt, Lars. "Spatio-Structural Granularity of Biological Material Entities." *BMC Bioinformatics* 11 (2010): 289.

Vogt, Lars, Peter Grobe, Björn Quast, and Thomas Bartolomaeus. "Accommodating Ontologies to Biological Reality—Top-Level Categories of Cumulative-Constitutively Organized Material Entities." *PLoS ONE* 7 (1) (2012): e30004.

Vogt, Lars, Peter Grobe, Björn Quast, and Thomas Bartolomaeus. "Fiat or Bona Fide Boundary—A Matter of Granular Perspective." *PLoS ONE* 7 (12) (2012): e48603.

Vogt, Lars, Peter Grobe, Björn Quast, and Thomas Bartolomaeus. "Top-Level Categories of Constitutively Organized Material Entities—Suggestions for a Formal Top-Level Ontology." *PLoS ONE* 6 (4) (2011): e18794. doi:10.1371/journal.pone.0018794.

Wakefield, Jerome C. "Biological Function and Dysfunction." In *Handbook of Evolutionary Psychology*, ed. David M. Buss, 878–902. New York: Wiley, 2005.

Zemach, Eddy. "Four Ontologies." *Journal of Philosophy* 23 (1970): 231–247.

Zhou, Yujiao, Bernardo Cuenca Grau, Ian Horrocks, Zhe Wu, and Jay Banerjee. "Making the Most of Your Triple Store: Query Answering in OWL 2 Using an RL Reasoner." In *Proceedings of the 22nd International Conference on World Wide Web (WWW 2013)*, ed. Ian Horrocks, 1569–1580. London: Elsevier, 2013.

Index

Printed in the United States
By Bookmasters